Cognitive Applications in Resistive Memories

Von der Fakultät für Elektrotechnik und Informationstechnik der Rheinisch-Westfälischen Technischen Hochschule Aachen zur Erlangung des akademischen Grades eines Doktors der Ingenieurwissenschaften genehmigte Dissertation

vorgelegt von

Diplom-Ingenieur Elektrotechnik
Lutz Stephan Nielen
aus Neuss

Berichter: Univ.-Prof. Dr.-Ing. R. Waser
Univ.-Prof. Dr.-Ing. T. Gemmeke

Tag der mündlichen Prüfung: 04.10.2019

Kurzfassung

Seit Jahrzehnten erfüllt die Entwicklung der Informationstechnik das berühmte Moore'sche Gesetz, welches besagt, dass sich die Komplexität von integrierten Schaltkreisen bei nahezu gleichbleibenden Kosten alle 12 bis 24 Monate verdoppelt. Daraus leitet sich unter anderem ab, dass auch die Leistung der Computertechnik der gleichen Entwicklung folgt. Lange konnte durch Miniaturisierung von Transistoren die Leistung von integrierten Schaltungen verbessert werden. Doch stößt diese Methode mittlerweile an physikalische Grenzen und es werden neue Technologien benötigt, um den bestehenden Trend aufrechterhalten zu können. Das gilt nicht nur für Digitalrechner, sondern auch für Informationsspeicher werden neue physikalische Prinzipien gesucht, die weiteres Skalierungspotential bieten. Nichtflüchtige resistive Schalter (auch als Memristor bekannt) bieten eine vielversprechende Alternative zu bestehenden ladungsbasierten Speichern wie der Flash-Technologie. Diese zweipoligen Bauelemente können aufgrund ihres sehr einfachen schichtweisen Aufbaus sehr gut skaliert werden und ermöglichen in einer passiven Integration den kleinstmöglichen Flächenverbrauch von $4F^2$. Resistive Schalter verfügen im Vergleich zu Flash-Speichern über wesentlich kürzere Schreib- und Zugriffszeiten bei geringerer Verlustleistung. Neben der Funktion als reines Speicherelement bieten sich insbesondere in einer komplementären Anordnung von zwei resistiven Schaltern Möglichkeiten für Logikanwendungen und neuronal inspirierte Methoden.

Im Rahmen dieser Dissertation werden zwei solcher weiterführenden Anwendungen für resistive Schalter näher beleuchtet. Zum einen werden Assoziativspeicher basierend auf zwei komplementären resistiven Schaltern, Complementary Resistive Switch (CRS), vorgestellt. Hier werden die gegenpolig verschalteten bipolaren resistiven Schalter so erweitert, dass durch den Speicherzustand eine Kapazitätsänderung erreicht werden kann. Diese Kapazitätsänderung kann ausgenutzt werden, um in passiven Speichermatrizen durch simultanes Vergleichen eines Suchworts mit dem gesamten Speicherinhalt die Ähnlichkeit der gespeicherten Wörter mit dem Suchwort zu bewerten. Diese in der Informationstheorie als Hamming-Distanz bezeichnete Ähnlichkeit ist ein Maß dafür, wie viele Bits zweier gleichlanger Datenwörter unterschiedlich sind. Da in dieser Technik der Speicher als großes Netzwerk von verschiedenen Kapazitäten aufgefasst wird, wird diese Technik als assoziatives kapazitives Netzwerk, Associative Capacitive Network (ACN), bezeichnet. Solche Assoziativspeicher werden in der Netzwerktechnik bereits verbreitet angewendet, um Adresstabellen in Routern effizient zu gestalten. Die hier vorgestellte Technik kann die Effizienz bezüglich des Energieverbrauchs pro Suche und, aufgrund ihrer Nichtflüchtigkeit, den Energieverbrauch im Standby wesentlich verbessern. Solche Netzwerke werden in dieser Arbeit ausführlich beschrieben, mithilfe von Kompaktmodellen in Simulationen evaluiert und technologisch als Hardware getestet.

Zum anderen wird eine Methode für passive Sortiernetzwerke vorgestellt, in der mithilfe von resistiven Schaltern verschiedene analoge Signale verglichen werden. Basierend auf der Theorie der Fuzzylogik werden die zwei grundlegenden Funktionen ‚Minimum‘ und ‚Maximum‘ mit resistiven Schaltern umgesetzt. Diese beiden Gatter wurden hergestellt, getestet und evaluiert. Mithilfe dieser beiden Grundfunktionen der Fuzzylogik kann ein Komparator aufgebaut werden, der das Kernelement eines analogen Sortiernetzwerks darstellt. Ein solches Sortiernetzwerk wird in Form von kompaktmodellbasierten Simulationen beschrieben und auf Realisierbarkeit und Verhalten hin untersucht.

IV

Abstract

For decades, the development in information technology has met Moore's famous Law, which states that the complexity of integrated circuits doubles every 12 to 24 months at a nearly constant cost. In consequence, the performance of computers follows the same trajectory. The performance of integrated circuits could be improved by the miniaturization of transistors. However, this method has reached physical limits now and new technologies are required to sustain the existing trend. This requirement for new concepts with wider scaling potential applies to digital computers, but also within the field of information storage. Nonvolatile resistive switches (also known as memristors) are a promising alternative to current charge-based memories like flash-technology. These two-terminal devices have high scaling potential due to their very simple layer structure and enable the smallest feasible area consumption of $4F^2$ by passive integration in crossbar arrays. The access time of resistive switches is much shorter compared to flash memories and they display lower power dissipation. In addition to the memory function, these elements offer opportunities for logic applications and brain-inspired neuromorphic methods. Especially the complementary assembly of two resistive switches facilitates novel beyond-von-Neumann concepts.

In this thesis two of these advanced applications for resistive switches will be highlighted. Firstly, content addressable memories based on two Complementary Resistive Switches (CRS) are presented. Here, the setup of the two anti-serially connected bipolar resistive switches enables a capacitance change depending on the stored memory state. This change in capacitance can be exploited to evaluate the similarity between a certain search pattern and the stored patterns within a passive memory array. This can be achieved by comparing the search pattern with the entire content of the memory simultaneously. This similarity is named Hamming Distance in information theory and is a measure for the number of different bits in two strings of the same length. Since the memory is considered a large network of different capacities in this technique, this concept is referred to as an Associative Capacitive Network (ACN). Such content addressable memories are widely used in network applications to efficiently design look-up tables for network routing. The technique presented here is capable of improving the energy consumption per search iteration and the standby-power significantly due to its nonvolatility. In this work ACNs are described in detail, they are evaluated by simulation with the help of compact models and they are technologically tested with hardware circuity.

Secondly, a method for passive sorting networks is presented, where resistive switches enable the comparison of variant analog signals. Based on the theory of fuzzy logic, the two fundamental fuzzy logic functions 'Minimum' and 'Maximum' can be realized using resistive switches. Both fuzzy logic gates were manufactured, tested and evaluated. These two basic functions of fuzzy logic allow the formation of a comparator, the core element of an analog sorting network. The feasibility of such a sorting network is investigated and described, supported by compact-model-based simulations.

VI

Acknowledgment

This thesis was written during my doctoral research at the Institut für Werkstoffe der Elektrotechnik II (IWE II) at the RWTH Aachen University. This dissertation would not have been possible without the support of many people who shall be mentioned in the following.

First and foremost, I would like to thank Prof. Dr. Rainer Waser, head of the institute, for his support and the opportunity to work within his inspiring research group for resistive switches for many years. I appreciate the great freedom and the high variety of interdisciplinary activities. Moreover, I would like to send my thanks to Prof. Dr. Tobias Gemmeke who kindly agreed to be the co-examiner of my doctoral work.

I would like to express my gratitude to Dr. Eike Linn for his guiding help, for his constant advice and support, during this work. The same holds Dr. Stefan Tappertzhofen, who helped me during this work with word and deeds. I also want to thank Dr. Omid Kavehai, who significantly contributed to the topic of this work.

To Martin Klimo and Ondrej Šuch I am grateful for informative and educational discussions about the field of fuzzy logic and sorting networks.

I thank Anne Simon for her excellent collaboration and the important computational support as well as Dr. Tobias Breuer his support with experimental data.

I would like to thank Dr. Stephan Menzel, Dr. Dirk Wouters and Dr. Ulrich Böttger who helped me with their advice and many fruitful discussions.

Many thanks go to all research assistant and student who contributed to this work with many hours of work in the lab or in the office. Special thanks go to Dr. Roland Rosezin, Susan Ohm, Joaquin Velasco Jitton, Nan Zhang, Kevin Strehlke, Carsten Beckmann. Furthermore, I want to thank Martina Heins, Udo Evertz, Thomas Pössinger, Dagmar Leisten, Daliborka Erdoglija, Petra Grewe, Gisela Wasse, Katharina Utens and Peter Roegels for administrative and technical support. I would also thank the electronics labs, namely Jochen Heiss and Hartmut Pütz, as well as the whole mechanical workshop under the lead of Karl-Heinz Stellmach.

Finally, I dearly thank my wife Inka Nielen for her dedicated and sacrificing support to finish this dissertation. I also thank Lena Müller, Aoibheann Rogers and Moritz Müller for proof-reading. I would like to thank my family and all friends for their support and their encouragement to this work.

VIII

List of Content

XII

Abbreviations

1D-1R	Single Diode / Single RRAM cell
1T-1R	Single Transistor / Single RRAM cell
AC	Alternating Current
ACN	Associative Capacitive Network
AD / ADC	Analog-to-Digital Converter
BE	Bottom Electrode
BL	Bit Line
BNC	Bayonet Neill–Concelman
CAM	Content Addressable Memory
CB-RAM	Conductive Bridge Random-Access Memory
CC	Current Compliance
CMOS	Complementary Metal Oxid Semiconductor
CPU	Central Processing Unit
CRS	Complementary Resistive Switch
DA / DAC	Digital-to-Analog Converter
DC	Direct Current
DIL	Dual In-Line Package
DRAM	Dynamic Random-Access Memory
DUT	Device Under Test
DVD	Digital Versatile/Video Disc
ECM	Electrochemical Metallization Mechanism
EEPROM	Electrically Erasable Programmable Read-Only Memory
FET	Field Effect Transistor
FPGA	Field Programmable Gate Array
GMR	Giant Magnetoresistance
GPU	Graphics Processing Unit

HD	Hamming Distance
HDD	Hard Disk Drive
HRS	High Resistive State
INV	Inversion Gate
IPSec	Internet Protocol Security
IRDS	International Roadmap for Devices and Systems™
LRS	Low Resistive State
LSB	Least Significant Bit
MAX	Maximum Gate
MIEC	Mixed Ionic-Electronic Conducting
MIM	Metal-Insulator-Metal
MIN	Minimum Gate
ML	Match Line
MOSFET	Metal Oxide Semiconductor Field Effect Transistor
MSB	Most Significant Bit
MSD	Mass Storage Device
MTJ	Magnetic Tunnel Junction
NAND	Not-AND
NDRO	Non-Destructive Readout
NOR	Not-OR
NVM	Non-Volatile Memory
PCM	Phase Change Memory/Material
PMC	Programmable Metallization Cell
RAM	Random-Access Memory
RBS	Rutherford Backscattering Spectroscopy
ReRAM	Redox-based Resistive Switching Random-Access Memory
RF	Radio Frequency
ROM	Read Only Memory
RRAM	Resistive Switching Random-Access Memory

SL	Search Line
SMD	Surface Mounted Device
SPICE	Simulation Program with Integrated Circuit Emphasis
SRAM	Static Random-Access Memory
SSD	Solid-State Drive
STT	Spin-Transfer Torque
TCAM	Ternary Content Addressable Memory
TCM	Thermochemical Mechanism
TE	Top Electrode
UV	Ultra-Violet
VCM	Valence Change Mechanism
VL	Voltage Line
WL	Word Line
WLDA	Winner-Loser-Distance-Amplification
WTA	Winner-Take-All
XNOR	Exclusive-Not-OR

Naming conventions

<u>ReRAM:</u>	Single switchable metal-insulator-metal stack	element
<u>CRS</u>	Stack of two anti-serially connected ReRAM elements	device
<u>NDRO</u>	Stack of two anti-serially connected ReRAM elements with significantly different element capacitances	(NDRO-) device
<u>ACN</u>	Assembly of two NDRO-cells which share one word line	(ACN-) cell

XVIII

1 Introduction

"Memory technologies will continue to drive the most aggressive pitch scaling and the highest transistor count. Memory technologies have been and will always be the drivers of Moore's Law" (2015's reports of the International Technology Roadmap of Semiconductors [1, 2]).

The upcoming trends in information technology follow irresistibly the predicted changes of the Internet of Things (IoT). More and more consumer electronic devices, from mobile phones and tablets to home entertainment and management electronics to vehicles and other items, are connected in large networks of software, sensors, actuators, computational power and memory. That requires enhanced performance of mobile electronics, thus memory techniques face a growing number of challenges. Those challenges are better performance in terms of speed, endurance and retention, lower energy consumption and cost efficiency. In accordance with Moore's law [2], the number of integrated devices (i.e. typically transistors) per chip increases exponentially with time. This was traditionally achieved by transistor scaling; however, this trend will end by reaching physical limits which will not allow further shrinking of integrated devices [3].

Redox-based resistive switches (ReRAM) are highly attractive devices for implementation of ultimately-scaled energy-efficient memories [4, 5]. Resistive switches offer memristive behavior, thus are also called memristive devices [6, 7] or, nowadays, memristors for short [8]. Besides non-volatile data storage, memristive devices offer intrinsic data-processing capabilities widening the scope of application far beyond simple memory. For example, computing operations for calculations were performed [9, 10], behavior of biological synapses was emulated [11] and circuits for realization of specific neuromorphic tasks were developed [12, 13]. In this work a hardware implementation enabling efficient pattern recognition tasks using advanced ReRAM devices is demonstrated. The utilized concept can be classified as an Associative Capacitive Networks (ACN) [14], which enables the implementation of Content Addressable Memories (CAM). Besides conventional state-of-the-art SRAM based CAMs [15, 16] which are limited by power consumption and cell area, several alternative implementations for Content Addressable Memories have been developed recently. Furthermore, there are optimization approaches using integrated DRAM cells [17] and approaches that include non-volatile techniques like floating gate MOS-FETs [18] or Spin-Torque-Transfer cells [19,

20] as well as principles based on phase change materials [20] and ferroelectric memories [21]. In addition, the use of ReRAM cells has been suggested [22-26]. However, in general all these concepts are not compatible to transistor-free $4F^2$ passive crossbar arrays or require a scalable bipolar selector [4, 27].

Figure 1.1 illustrates the evolution from the biological neuron to the artificial synapse [28] and the here presented associative capacitive network. A key feature of RRAM that is also highlighted in the IRDS [4] is the potential to integrate an artificial neural network. That is usually achieved by the multistate property of RRAM cells that are gradually changed by voltage spikes. However, it indicates the capabilities to solve computational tasks within the memory itself [29].

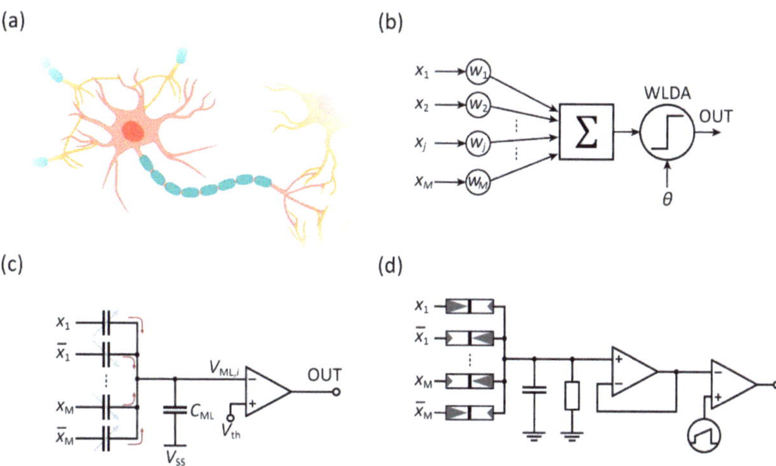

Figure 1.1: Neuromorphic inspiration – (a) Typical illustration of a neuron. The cell body (soma) receives information signals by various tree-like extensions, the dendrites. Information is processed in the soma and transferred by the single long axon. The axon is shielded by myelin sheaths to isolate the electrical signal, and it can reach a length up to 3 m. The axon ends up in several branches, the axon terminals, which are connected to the dendrites and somas of other neurons by synapses. (b) The artificial neuron consists of a simplification of all the important neuronal functions. The influence of M input signals is defined by the individual weight w_j. The transmission function Σ summarizes the weighted input. Subsequently, the output result of the neuron, OUT, is defined by the activation function. With a certain threshold value θ a winner-loser-distance-amplification (WLDA) is defined. (c) An associative capacitive network realizes the weighting of the inputs by a capacitive voltage divider. The activation function can be implemented by a CMOS comparator. (d) The here demonstrated ACN utilizes complementary resistive switches (CRS) to adjust the capacitance for the weighting. The threshold value is defined by voltage ramp which allows a validation of the result in the time domain.

2 Fundamentals

In computer science, a memory device is considered as an electronical part which stores information, i.e. data. As this work is about one specific type of memory, firstly some general properties of memory are described. Every class of memory needs to fulfill some fundamental requirements. The basic memory element needs to have at least two distinguishable states. As state-of-the-art computers are binary these two states are sufficient. These states must be detectable by an electronic sensing process. This process is called the READ process. Furthermore, there must be a mechanism to set each memory element into a specific state. If this process is only achievable once, e.g. during manufacturing of the memory, the memory is classified as read only memory (ROM). Nowadays, memory has usually the demand of a WRITE process. Hence, there must be a process to change the state of the memory element multiple times. That requires a SET process, which sets the element from its initial state (logic '0') to the other memory state (logic '1'). The process which recovers the initial state is called the RESET. Of course, it is just a matter of definition which state is the logic '0', and which is the logic '1'. These read & write memories built the scope of today's information memory devices and storage system.

In the following different classes of memory, the fundamental memory principle of the resistive switch, and future opportunities for resistive switches are described.

2.1 Classes of Memory

Electronic mass storages are grouped into two major systems, working memory and mass storage devices (MSD). In traditional mass storage devices the information is usually sequentially accessible and access depends on the physical location on the data medium, so it is a relatively slow technique. The key feature of random access memory (RAM) is parallel access which significantly decreases the access time. That is realized by assembling the single information storage elements in a matrix. In matrices information can be accessed block-wise or individually at any time. Working memory is usually realized by a RAM implementation. That is also the reason why working memory is also often just called RAM. Traditionally, working memory is especially used to operate close to CPUs, whereas MSDs have rather the function of long time data storage. That causes a certain hierarchy of memory classes according to the rules of the von Neumann architecture [30].

In summary, MSDs and RAMs vary in the following basic principles. In MSDs the information is stored in a specific contiguous medium and can be accessed by an individual unit, typically a Read/Write (R/W) head. Medium and R/W head need to be mutually aligned to address information in the memory. The exchange mechanism can have manifold physical concepts, e.g. magnetic (hard disk drives, tape records), optical (compact disk, DVD, holographic), mechanical (scanning probe storage), or others. The task of the access unit is to convert and exchange information from the medium to an electric signal. In RAMs conductor lines are assembled to a matrix with the storage elements at its intersecting nodes. By selecting the right combination of column and row lines, each node can be individually addressed. The selection needs to take place by an electrical signal, i.e. applied voltage or current. The conversion from an electrical signal to the individual storage medium needs to happen at the storage element itself, which makes some physically complex methods unfeasible. [31]

In the recent past, a concept for computers was developed with three basic memory levels. Figure 2.1 shows the different levels in a pyramidical hierarchy. The highest level is the CPU cache. It is SRAM based and stores information which directly serves the calculation in the CPU. The major demand to this memory is speed and compatibility to the arithmetic unit. SRAM can be perfectly integrated to CPUs. Results and tasks running program which are not immediately needed for further calculations are stored in the main memory, i.e. working memory or RAM. It still has a certain requirement to access speed, but factors like memory size and power consumption become more important to this level of memory. Recently, dynamic RAM (DRAM) which stores electric charge on capacitors has become the means of choice to realize main memory. The third level of computer memory is the mass storage memory for long time information storage. The hard disc drive (HDD) was and still is the widest spread solution in consumer electronics for this task. It is based on the giant magnetoresistance (GMR) effect, i.e. a magnetic field effect, discovered in 1988 by P. Grünberg [32] and A. Fert [33] and belongs to the class of MSDs. Due to the nature of sequential data access in MSDs, writing and reading times are rather long. However, in the last years it has turned out, that access time on HDDs has become more and more a bottleneck on the way to faster and more powerful computers. A new technology, the solid-state drive (SSD), has extensively replaced former hard drives. The storage mechanism in State-of-the-art SSDs is Flash-EEPROMs (short FLASH), which are based on the physical concept of floating gate MOSFETs. Charge is trapped in a conductive element within the insulating oxide layer of a MOSFET [34]. As each memory element is a single transistor, this technique is perfect for integration in a matrix. That could significantly improve the access time of mass storage.

In contrast to the high-level memory classes SRAM and DRAM, mass storage devices must be non-volatile. That means that stored data remains in the memory for years without any refresh procedures, i.e. without power supply. This is a crucial difference between these classes.

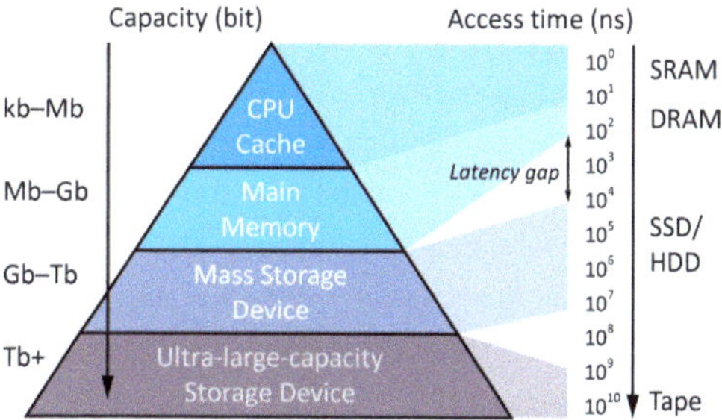

Figure 2.1: Hierarchy of computer memory classes – the typical storage size increases from top to bottom. Labels within the pyramid describe the function of each memory class. At the right axis their corresponding technology is matched to the typical access times. In current technologies there is a latency gap between DRAM and SSDs which motivates the need of a new class of memory. Adapted from [35, 36].

Despite this technology leap of mass storage, flash memory still has capacity for improvement. Due to the fundamental writing process in flash memory, which is based on a tunneling effect through the oxide layer, the writing speed of a single memory element is still comparably low, i.e. in the range of milliseconds. There is a considerable latency gap between the main memory and storage devices (cf. Figure 2.1). This writing process also requires a voltage of typically 5 V or even higher voltages, which is rather high for common CMOS integration. This voltage demand requires a noticeable power consumption. In fact, around 30% of the energy of modern mobile communication electronics is spent on memory.

Like all microelectronic architectures, FLASH memory concepts were continuously improved by scaling for better power consumption, higher speed/performance, and lower costs following Moore's law [2].

2.2   Fundamentals of Resistive Switching

The basic element which is the fundament of all studies in this work is the resistive switch. The class of resistive switches summarizes information storage devices which are based on a change of its electrical resistance. There are manifold physical effects that can cause a change of resistance in its internal structure. Figure 2.2 gives an overview of various resistive switching effects. In general, they have in common that a certain part of the device changes its resistance. This resistance can be accessed and readout by two terminals. For some technologies there are further terminals needed to change the resistance, e.g. magnetoresistive memory core cells exhibit up to five terminals. However, the promising STT-RAM has also two terminals only. In this work though, the redox-based resistive switches (ReRAM) is the fundamentally discussed element. It can be simply described as a metal-insulator-metal (MIM) structure. That means that two conductive elements, which represent the mentioned two terminals, sandwich a non-conducting material. Accordingly, the write and read process is performed by the same terminals. Appling voltage to these devices causes a change of the electrical conductivity in the insulator material. In practice, this causes a hysteretic resistance-to-voltage diagram. Hysteretic characteristics are typical for memories, since in the zero point of the x-parameter the characteristic defines two values for the y-parameter.

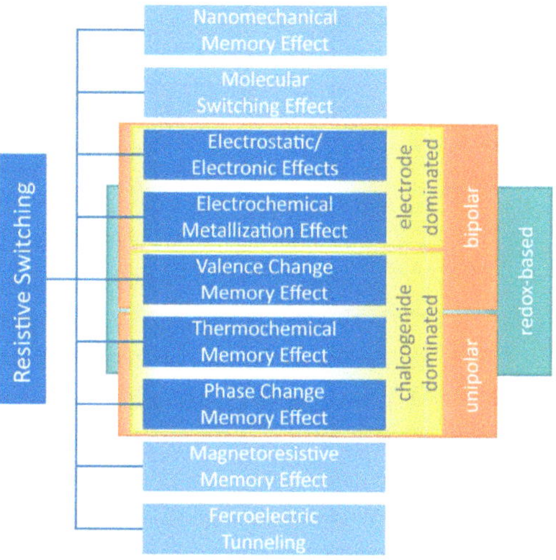

Figure 2.2: Classification of resistive switching mechanisms for memory applications. A non-volatile switching effect is an essential requirement for permanent mass storage uses. Key criteria for classification are the distinction between uni- and bipolar switching and between chalcogenide and electrode dominated switching effects. Redox-based resistive switches (ReRAM) is the category for switching mechanisms based on electrochemical metallization (ECM), valance change (VCM) and thermochemical (TCM) effects. Reworked from [5].

2.2.1 General Functions

Depending of the physical/chemical effects and the electrical behavior during the switching, two essential classes of ReRAM switching are observed. So these elements can exhibit unipolar or bipolar switching. Figure 2.3 shows the characteristic IV-curves of these two types. Both types reveal SET and RESET mechanisms – a fundamental requirement for reprogrammable memory, as described earlier. Unipolar switching can be performed with only one voltage polarity. Here, the SET event occurs at a higher voltage than the RESET effect. As the RESET at unipolar switching is triggered by the significantly higher current, a current compliance usually ensures that the RESET is not immediately initiated after switching (cf. Figure 2.3a). In contrast to this operation mode bipolar switching requires two different polarities of voltage to perform a SET or a RESET, respectively (cf. Figure 2.3b). Here, a current compliance is often used to adjust the resistance after the SET procedure. The common convention is that the SET process sets the element from the high resistive state (HRS) or OFF state, respectively, to the low resistive state (LRS) or ON state, respectively.

Depending on the exact choice of materials, different physical/chemical effects cause the conductivity change within the resistive switch. As illustrated in Figure 2.2, the class of redox-based resistive switches (ReRAM) can be distinguished in three major subclasses: The electrochemical metallization (ECM), the valence change effect (VCM) and the thermochemical effect (TCM). In the following, the fundamental physics of ECM and VCM are described. TCM is based on an effect of thermally induced variations in the stoichiometry and redox reactions [37]. As TCM is a unipolar switching effect it has no importance to this work. The state of art and physics of ReRAMs have been studied and reviewed by Valov et al. [38-42], Waser et al. [5, 43], and Ielmini et al. [37].

Nanoelectronics integration concepts for resistive switches into state-of-the-art computer chips are numerous. Extensive studies were outlined by Yang et al. [44] and Pan et al. [45].

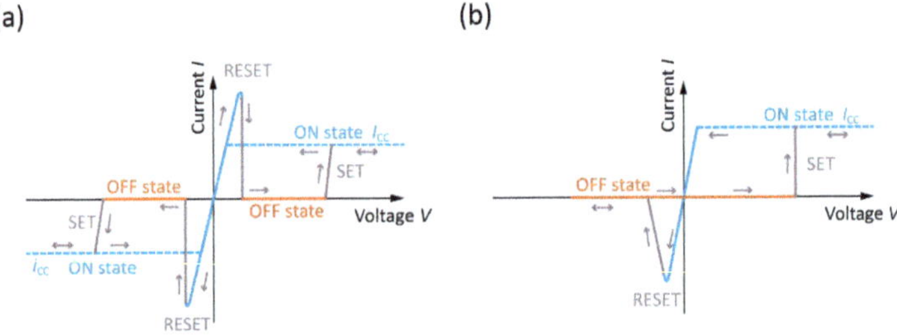

Figure 2.3: Basic operations of resistive switching memory cells classified by the operational voltage regime. The characteristic current–voltage diagram is recorded for a triangular shaped voltage signal. Blue and orange color indicate the *I–V* characteristic for LRS and HRS, respectively. A current compliance is commonly used for resistive switching which is abbreviated by CC. When the compliance is activated during the SET process the applied voltage drops immediately to a lower value which is indicated by the dashed lines. (a) Unipolar switching: SET and RESET events occur for the same voltage polarity, whereas the SET voltage is always higher than the RESET voltage (here, both polarities are shown). (b) Bipolar switching: The SET process is triggered by a positive voltage while the RESET operation requires a negative voltage. Redrawn from [5].

2.2.2 ECM

The effect of electrochemical metallization memories (ECM), also known as Conductive Bridge (CB-RAM) or Programmable Metallization Cells (PMC), can be described with the help of Figure 2.4. The cell consists of an electrochemically active electrode, such as Silver or Copper, a chemically inert counter electrode, such as Platinum, Iridium, Gold or Tungsten, and a thin film of an chalcogenide or insulating oxide layer, like $GeSe_x$ [46],TiO_2 [47] or SiO_2 [48, 49] in this example. The transition between HRS and LRS is caused by the formation and rupture of a nanoscale conductive filament [39]. In Figure 2.4, a quasi-static triangular voltage signal is used for the I–V diagram. Starting at the initial high resistive state (A), when no filament has developed yet, the current slightly increases with rising voltage. Here, the current is mainly caused by the leakage current of the solid electrolyte. During the SET process a positive voltage is applied at the active electrode and an anodic oxidation of the active electrode metal injects positively charged metal ions into the insulator (B). Further increase of the voltage, and consequently the driving electric field, leads to a reduction of the positive metal ions at the inert counter electrode. That results in the formation of a conductive filament of metal atoms that grows towards the active electrode. When the filament bridges the path between the two electrodes, the LRS is achieved and the current rises instantly (D). A current compliance usually prevents the filament from accidental damage by two high currents, stops the formation process and limits the ON resistance.

Now, the I–V curve exhibits a very steep but linear slope around the zero point. This LRS state is non-volatile, i.e. the state remains even when no voltage is applied. The HRS can only be recovered by applying a significantly high negative voltage to initiate the RESET operation. Now, the filament is dissolved by the same electrochemical oxidation and reduction processes that also led to the growth of the filament (E). The existence of a filament based mechanism is motivated by the observation that the process is independent of the electrode area [50].

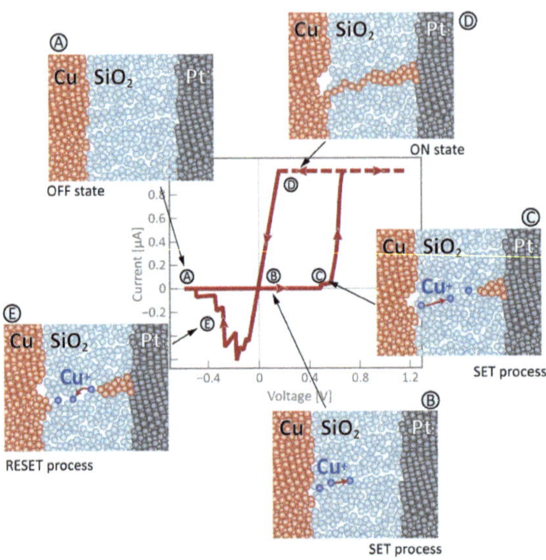

Figure 2.4: General principle of resistive switching based on the electrochemical metallization effect (ECM) during a voltage sweep. In the current–voltage diagram characteristic events are explained by pictograms of the MIM stack. In this Cu/SiO$_2$/Pt setup positive Copper ions are dissolved out of the reactive electrode and propagate through the oxide layer to build a filament on the inert electrode. The insets show the pristine (A) HRS (B)–(C) SET process (D) LRS and (E) RESET process. Reworked from [39].

2.2.3 VCM

In contrast to ECM based resistive switching, the process of VCM switching (also called Anion-type RRAM) is dominated by a mixed ionic-electronic conducting (MIEC) chalcogenide layer. An asymmetry in the material needs to be gained for bipolar switching. That can either be achieved by different electrode material (like in the following example) or by an initial electroforming procedure [5]. Many different variations of VCM-type switching are known [36].

In this example, there is an active interface (active electrode) where the switching occurs. The other terminal is an ohmic counter electrode. Here, the mechanism is exemplarily described for the case of Pt/ZrO$_x$/Zr stack. Figure 2.5 shows a typical *I–V* diagram generated by a triangular voltage signal. Assuming that the cell is in the HRS after an initial forming procedure, the tip of the *n*-conducting MIEC oxide filament (called plug) and a potential barrier between the tip of the filament and the electrode (called disc) build the elementary switching part. This situation is shown in pictogram (A). Applying a negative voltage initiates oxygen vacancies to move from the plug into the disc (B). The local reduction of oxygen vacancies in the disc causes a significant decrease of the potential barrier. That can be observed in the increase of conductivity, thus in an increase of current (C). Decreasing the applied voltage reveals that the cell's resistance was set to the LRS. When a positive voltage is applied, oxygen vacancies are removed from the disc back to the plug where a local re-oxidation resets the initial HRS.

VCM-type resistive switching has been intensively reviewed by Waser et al. [5]. The variety of different VCM type switching effect includes binary oxides like TiO_2 [51-54], HfO2 [55, 56] or Ta_2O_5 [57, 58], perovskite oxides like $SrTiO_3$ [59, 60], as well as complex and doped oxides [45, 61-62].

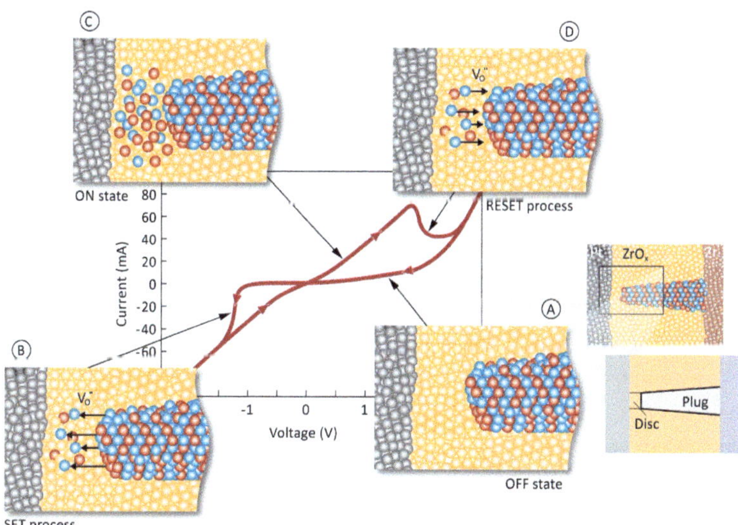

Figure 2.5: General principle of valence change memory (VCM). In this example the current–voltage characteristic of a $Pt/ZrO_x/Zr$, which is stimulated by a triangular signal, is marked with insets for the crucial events of the switching process. Here, the Platinum represents the active electrode and the Zirconium the ohmic one. The pictograms show only a cutout of the part with the interface and the tip of the conductive filament. The dark red spheres indicate the Zr ions in a lower valence state and blue spheres stand for oxygen vacancies, the only mobile particles in this layer. (A) HRS (B) SET process (C) LRS (D) RESET process. Reworked from [63].

2.2.4 Switching Kinetics

All types of ReRAM cells exhibit strongly nonlinear switching kinetics. This effect was intensively studied at the example of $SrTiO_3$ [64, 65]. This nonlinearity results in the effect that very short writing pulses, which are demanded in emerging memories, require essentially higher writing voltages. Figure 2b in [64] gives a conclusive overview about various pulse lengths in relation to the applied voltage. This nonlinear voltage acceleration of the kinetics of many orders of magnitude is also known as the voltage–time dilemma. The curves which are shown in Figure 2.4 and Figure 2.5 are not affected by this effect, since the quasistatic voltage sweep can be considered as a train of relatively slow voltage pulses (range of hundreds of milliseconds). The nonlinear effect can most likely be ascribed to electric-field-enhanced ion-hopping mobility [66, 67]. Also, the effect of electric-field enhanced recombination/generation of oxygen vacancies can be a potential origin for VCM-type resistive switching kinetics [68].

2.2.5 ECM vs. VCM

The mechanisms of the resistive switch classes that play a role in this work are described in Section 2.2.2 and Section 2.2.3. This part points out some important differences of both effects that are interesting for the switching characteristics in this work.

Both effects are capable of high ON to OFF ratios, i.e. the high resistive state exhibits resistances that are in some order of magnitude higher than the low resistive state. ON to OFF ratios in the range of 10^6 have been shown for VCM [69] and values $> 10^7$ for ECM [70]. However, due to its general mechanism this feature is rather typical for ECM cells. The LRS of ECM cells is dominated by the metallic properties of the filament, therefore it has a linear ohmic behavior whereas the HRS is dominated by the insulating oxide. This state can exhibit non-linearities due to the conductance mechanism of oxides. VCM typically exhibits strong non-linearities in the HRS. The LRS is predominantly linear. That makes simplifications and integration in more complex circuits challenging.

In general, the polarity of these bipolar switches can be adapted to memory applications. The definition of which state is a logical '0' or '1' is permutable. By taking a closer look at the formation of a filament there is a difference between ECM and VCM. A positive voltage on an ECM cell forms a filament from the cathode to the anode. Therefore, the SET voltage V_{SET} is positive as illustrated in Figure 2.4. In contrast to that, filaments in VCM cells are formed from the cathode to the anode by applying a negative voltage. This difference needs to be taken in mind for the individual scenarios of the complementary resistive switches which are presented in the following chapters.

2.3 Complementary Resistive Switch

The anti-serial assembly of two bipolar ReRAM elements in a Complementary Resistive Switch (CRS) was initially developed to overcome the so-called sneak path obstacle in passive crossbar arrays [27].

One major issue regarding passive crossbar arrays of resistive switching devices is the accurate addressing of single memory elements. Because of its passive nature, ReRAM arrays are large networks of highly interconnected resistors. In crossbar arrays a single element is addressed by applying voltage on a defined vertical and horizontal crossbar line, i.e. bit and word line. So, the voltage is applied to the whole conducting element. By that a current needs to flow through the interlinked neighboring memory elements. Figure 2.6 indicates the effect by a unified 3×3 array cut-out.

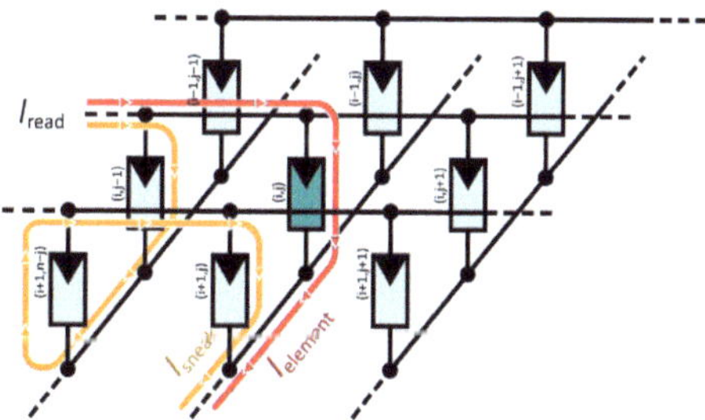

Figure 2.6: Illustration of the so-called sneak path problem. The figure shows the cut-out of a passive $m \times n$ array at position (i,j). The current $I_{element}$ through the element which is addressed is indicated by the red current path, whereas the yellow current path I_{sneak} demonstrates one possible current path through three neighboring memory elements. These parasitic currents will occur in all parasitic memory cells and interfere with the measurement current I_{read}.

Several options are possible to overcome this obstacle. All implementations involve a certain kind of selector device which disables the impact of neighboring elements. Currently the research is focused on the following candidates for these selector devices, (a) intrinsic nonlinearity of the resistive switch (b) extrinsic nonlinearity, i.e. diodes – 1D-1R cells (c) transistors – 1T-1R cells and (d) complementary resistive switches. Figure 2.7 illustrates these solutions. For options (a) and (b) only limited material systems can be considered, so option (c) is, until now, the most pursued candidate. Nevertheless, it involves the drawbacks of an active device, comparable high area demand and a second address line per column or row. The approach of a CRS device can overcome these disadvantages.

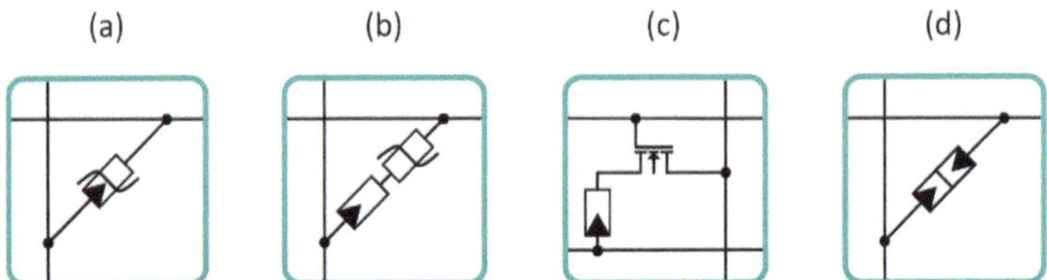

Figure 2.7: List of selector devices – (a) Intrinsic nonlinearity of the resistive switch, (b) 1D-1R cells, extrinsic nonlinearity realized by diodes for example (c) transistors – 1T-1R cells and (d) complementary resistive switches. Adapted from [71].

The main advantage of CRS is that in each logic state the overall device resistance is high, and thus the influence of neighboring memory cells is significantly reduced. Figure 2.8: shows the *I-V* characteristic of CRS devices.

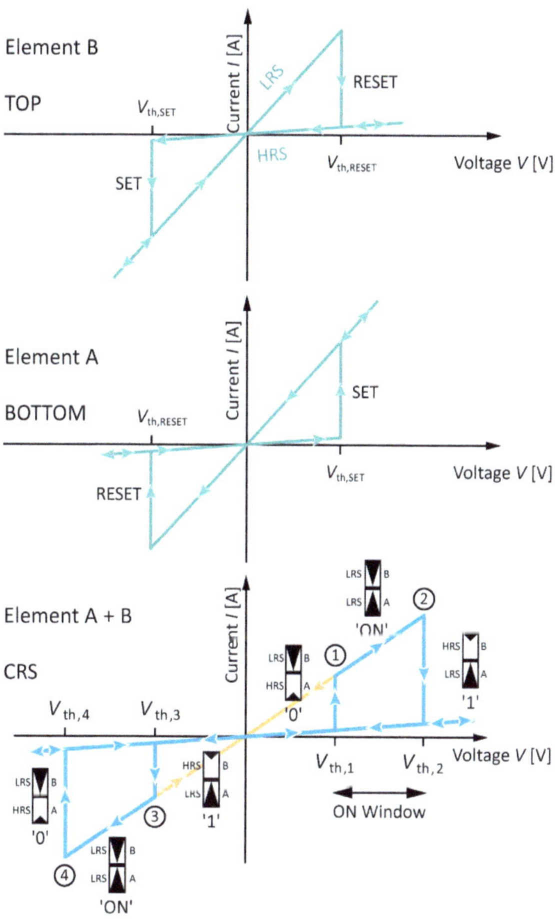

Figure 2.8: CRS *I-V* curve – the two top graphs show a simplified RRAM characteristic of two individual resistive switches. The second graph stands for Element A whereas the upper graph represents Element B which is assembled inversely and therefore exhibits a mirror-symmetric characteristic. Taking both characteristics together and considering the voltage divider properties of the stack results in the lower curve. The switching events ①–④ are explained in the text in more detail. Adapted from [27].

2.3.1 The principle

The setup of a complementary resistive switch consists of two anti-serially connected RRAM elements. That means that either both inert or both active electrodes are directly connected to each other. In terms of technology and fabrication that can be achieved by building a stack of two resistive switches in reversed order. The new device has only one top electrode (TE) and one bottom electrode (BE). Referring to the single resistive switch, both CRS electrodes are either manufactured from the inert or the active electrode metal.

To understand the principle of CRS, it is useful to simplify the I-V characteristic of a single resistive switch.

- First, it is assumed that the element has only two linear states, namely a low resistive state (LRS) and a high resistive state (HRS). Any transient or intermediate states as well as nonlinear effects, which occur in real ReRAM element, are neglected.
- Second, the resistance ratio of high and low resistive states is very high. So, it holds $R_{HRS} \gg R_{LRS}$.
- Third, the threshold voltage for both switching events is equal according to amount. In other words, $V_{th,SET} = -V_{th,RESET}$.
- Fourth, the two ReRAM elements of a CRS cell reveal the same switching characteristics in terms of resistance states and threshold voltages. The resulting I-V characteristic is sketched in Figure 2.8a and respectively, the reversed characteristic for the anti-serial element in Figure 2.8b.

With these simplifications the general switching properties of a complementary resistive switch can be explained. The CRS-state HRS/HRS cannot occur during normal CRS switching cycles. Depending on the forming procedure of CRS cells, different initial CRS states can be assumed. For best explanation, here the LRS/HRS state is assumed to be the initial state. That means that Element A is low resistive and Element B is high resistive:

①   Since the whole CRS stack represents a voltage divider with two different resistors, almost the complete applied voltage V_{in} drops on Element B, i.e. $V_{in} \approx V_B$. The voltage at Element A can be neglected. When the applied voltage is increased to the threshold level $V_{in} = V_{th,1} = |V_{th,SET}|$, the voltage at Element B causes the switching event which sets the element to LRS.

L/H
↓
L/L

②   Now, the applied voltage V_{in} is evenly divided between the two resistive elements. Accordingly, it holds $V_{in} = 2 \cdot V_A = 2 \cdot V_B$. When the applied voltage is further increased to $V_{in} = V_{th,2} = |2 \cdot V_{th,RESET}|$, the voltage which drops at Element A reaches the threshold voltage for the reset event. Thus, this element is switched to HRS. Theoretically, Element B also reaches the threshold voltage $V_{th,SET}$ at this moment. However, it is already in LRS, so the state does not change. Further voltage increment does not cause any switching events.

L/L
↓
H/L

③   Now negative applied voltages are considered with the previous switching history. The states of both elements have changed compared to the initial states. The overall CRS state is now HRS/LRS. Like in the beginning, the voltage division has become asymmetrical again and nearly all the applied voltage drops at the high resistive Element A, i.e. $V_{in} \approx V_A$. When the applied voltage is decreased to the negative voltage $V_{in} = V_{th,3} = |V_{th,SET}|$, Element A switches to the LRS.

H/L
↓
L/L

④   Again, both elements are in LRS and the applied voltage is divided evenly between both resistive elements. Further decreasing of the applied voltage leads to the switching event at $V_{in} = V_{th,4} = |2 \cdot V_{th,RESET}|$ which sets Element B in HRS. Element A is not further influenced by decreasing the voltage since it is already in LRS.
The initial CRS state is restored.

L/L
↓
L/H

This qualitatively derived *I-V* sweep reveals that the CRS stack can only generate three combinations of individual element states. That is LRS/LRS, HRS/LRS and LRS/HRS. The two combinations, HRS/LRS and LRS/HRS, can only be set up when one defined threshold was exceeded. That happens for the HRS/LRS state when $V_{th,2}$ is applied and for the combination LRS/HRS when $V_{th,4}$ is reached. So, these states are unambiguous and therefore suitable for information storage. In the following the combination HRS/LRS is defined as the logical '1' state, i.e. Element A is in HRS and Element B is in LRS. Vice versa the combination LRS/HRS stands for a logical '0'.

The state LRS/LRS appears two times in the cycle. That is always the case when the threshold voltage of the high resistive element is reached. With the initially assumed simplifications, the previous CRS state does not influence the behavior of the LRS/LRS combination. Thus, both appearances of the LRS/LRS state are not distinguishable. That makes them unusable for information storage. In the following this combination is also named 'ON'-state and is only considered as a transition state in valid CRS-operations that occurs during the switching process. However, one needs to keep in mind that the state can appear and is stable for $V_{in} = 0$ V, when switching voltages are not high enough to exceed $V_{th,2}$ or $V_{th,4}$, respectively. In Figure 2.8 this state is illustrated by the yellow curve. As the stack history has no impact on this 'ON'-state, it can be translated to both stable logic states, i.e. logical '1' and logical '0'. Table 2.1 gives an overview of the possible CRS states.

Name	**Element A**	**Element B**	Previous state	Threshold voltage
'1'	HRS	LRS	'ON'	$\uparrow V_{th,2}$
'0'	LRS	HRS	'ON'	$\downarrow V_{th,4}$
'ON'	LRS	LRS	'0' '1'	$\uparrow V_{th,1}$ $\downarrow V_{th,3}$
'H'	HRS	HRS	—	—

Table 2.1: Truth table for CRS devices – The valid states '1' and '0' are a combination of HRS and LRS for Element A and B. The 'ON' state is just a transition state from '1' to '0' and vice versa. The HRS/HRS state which is abbreviated with 'H' is a forbidden state in proper CRS operations. However, it is usually the initial state of CRS stacks after fabrication.

Until now the combination HRS/HRS was not discussed. As shown in Figure 2.8, this state cannot be gained by applying voltage to the CRS cell. Every polarity leads to one of the three CRS combinations in which at least one of the elements is low resistive. So, the HRS/HRS state does not need to be further considered for CRS operations. However, this state can be a real combination for two pristine ReRAM elements after fabrication. Thus, possible situations for this state need to be considered.

Two scenarios to change this state are conceivable. Any applied voltage is evenly divided between both elements. In Figure 2.9 the red curve shows the electric current which flows through the HRS/HRS cell. As the resistance is approximately double that of any HRS/LRS state, the current needs to be about half of the current that flows in the normal operations. Either Element A or Element B is switched to the LRS. Since the individual threshold is reached at $V_{in} = |2 \cdot V_{th,SET}|$, Element A switches at $V_{in} = V_{th,4}$ and Element B switches at $V_{in} = V_{th,2}$. Related to the nomenclature of Figure 2.9, the transition is either $(0) \rightarrow (2')$ to reach CRS state '1' or $(0) \rightarrow (4')$ for state '0'. Once the HRS/HRS state is left, it is not possible to switch back to this state unless a voltage is applied to the middle electrode.

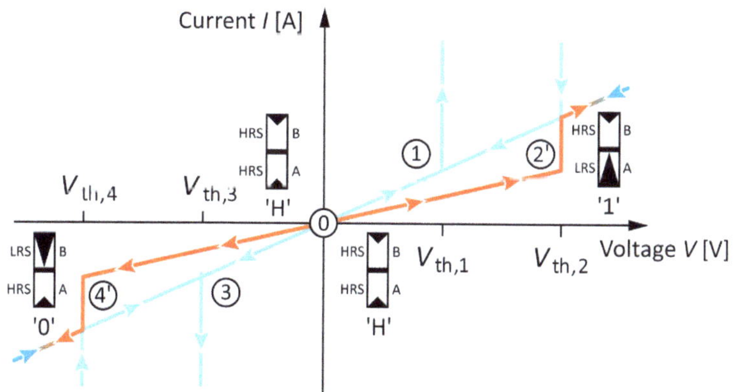

Figure 2.9: CRS *I-V* curve for HRS/HRS – The graph is a cut-out of the lowest graph in Figure 2.8. It is assumed that it behaves like the LRS/LRS voltage divider with a higher overall resistance. Thus, the switching to the valid logic states also appears at $V_{th,2}$ or $V_{th,4}$. Position ⓪ is considered as the absolute pristine state. From there, the device is either set by switching event ②' or ④'.

As a summary of the previously discussed switching mechanisms, Figure 2.10 gives an overview of previously described transitions. It shows the valid CRS-operations (blue cycle) the initializing process (red lines) and the possible transitions from the 'ON'-state to its previous state (yellow lines). The 'ON'-state is represented twice here, to show that it can be obtained by two different previous states, even if it is not intrinsically determinable. Both states are exchangeable.

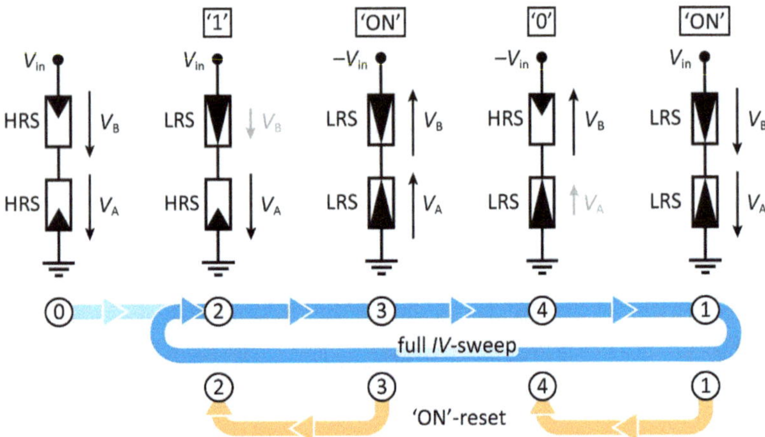

Figure 2.10: CRS state transitions – the diagram gives an overview of possible transitions. The blue circle represents the valid CRS-operations which is also the case in the *I-V* sweep of Figure 2.8. The red arrows show the initial set process from the pristine CRS-device to one of the valid logic states. The yellow arrows indicate the possible switching from the 'ON'-state to the opposite valid logic state. As the device is switched back to the state it had before it was set to the 'ON'-state, it is here labeled with the tag 'ON'-reset.

2.3.2 Readout

For application one important property of a memory device is the possibility of information readout. If information can be stored in a medium but there is no physical method to retrieve this information, it has no use in information technology.

The proposed CRS device sets a certain difficulty to this requirement, as unlike to single RRAM cells, the stored resistance cannot be measured with a small readout voltage around 0 V. From "outside" both logical states reveal a high resistance. For that reason, one needs to find an alternative method for the readout. The initially proposed method is realized by a destructive readout. Therefore, a voltage which exceeds the write threshold (i.e. $V_{th,2}$ or $V_{th,4}$) is applied to the cell. That causes the device to switch to the intended logical state. However, depending of the previous logical state, the device needs to pass through the transient 'ON'-state or keeps its state. Like in the I–V curve illustrated (cf. Figure 2.8) this 'ON'-state exhibits a significantly lower resistance. This fact can be exploited by two scenarios. Either a voltage that switches the CRS cell into the 'ON' window can be applied (Level Read) or a current spike is detected by a usual write operation (Spike Read) like Figure 2.11 illustrates with two exemplary timing diagrams. The Level Read is maybe more intuitive as the resistance is measured directly by applying a voltage which lies within this 'ON' window (cf. Figure 2.11a). However, the major drawback is the requirement of this third voltage level (besides $+V$ and $-V$) which needs to fulfill $V_{th,1} < V_{read} < V_{th,2}$. Especially for asymmetrical RRAMs where $|V_{SET}| \gg |V_{RESET}|$ this 'ON' window can become very narrow and thus the window becomes difficult to be controlled. A series resistor could widen this window during readout. In an integrated CRS array this resistor does not need to be integrated in every cross-point junction but could be externally added during the readout. This approach with a series resistor can be also very useful for studies of CRS behavior [72].

If the write voltage $+V$ is directly applied, a state change causes a higher current flow than for the case where the device remains in its previous state. This current flow lasts only for the time in which the cells are switching and the charge on the parasitic capacitance of the cells is reloaded. Accordingly, in the time diagram only a short current spike is detectable (hence, the name Spike Read, cf. Figure 2.11b). This spike can be detected for very short voltage pulses, even when the 'ON' window is not explicitly activated. Depending on the switching kinetics of the utilized RRAM elements and the parasitic cell capacitance this spike varies in height and length.

It is evident that this destructive readout has the need of a write-back operation after read. This subsequent write-back is not needed after each read operation but only for read operations that switch the cell, hence when the 'ON' window is measured for Level Read or current spike is detected for Spike Read (cf. step ③ and ④ in Figure 2.11). That means that in practice only every second read operation needs a write-back operation. Thus, the demanded time for read operations is 1.5 clock cycles in average.

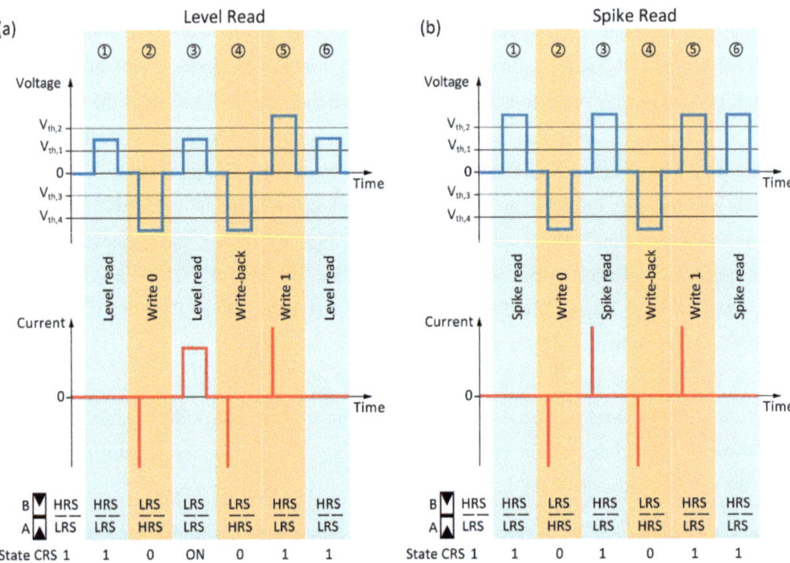

Figure 2.11: Timing diagram for conventional CRS readout – two general concepts can be applied: (a) Level Read and (b) Spike Read. In both concepts write operations are performed by applying voltage pulses that exceed the threshold voltages $V_{th,2}$ and $V_{th,4}$. A write operation to state '1' is performed by applying a negative voltage pulse ② and to state '0' by applying a positive voltage pulse ⑤. In this example three read operations are performed, in step ①, ③ and ⑥. It is noticeable that for (a) Level Read a voltage between $V_{th,1}$ and $V_{th,2}$ is used whereas for (b) Spike Read a usual write voltage ($V > V_{th,2}$) is applied. The result is noticeable in the current diagram. In step ③ a continuous current is flowing for the period of the pulse for Level Read (a) and just a current spike is detected for Spike Read (b). In the other read steps ① and ⑥ no current flow is measured since the CRS cell does not change its state. Accordingly, only after step ③ a write-back operation is required. This is done in step ④. The currents for writing do not differ for both cases. Adapted from [73].

2.3.3 Integration

The predicted behavior of CRS cells could be demonstrated in many experiments including VCM type and ECM type ReRAM elements.

The principle readout by current flow for short pulses have been demonstrated and characterized for different material compositions [58, 72, 74]. Fabrication and feasibility of crossbar structures have been demonstrated on the microscale [75, 76] and nanoscale. Here, an exemplary result based on Pt/SiO₂/Cu ECM cells is shown in Figure 2.12 [75, 76]. Figure 2.12a and b show the I–V curve of a 2×2 μm^2 crossbar integration with two single RRAM elements that share a common bottom electrode. This approach was chosen to ensure most similar properties of each single RRAM element. Furthermore this method reduces process steps during fabrication [75, 77]. Figure 2.12c shows the I–V characteristics of a vertically integrated CRS device consisting of the material stack Pt/SiO₂/Cu/Pt/Cu/SiO₂/Ti/Pt. This stack has an accessible middle electrode represented by the central Pt-layer. That allows monitoring of the voltage on each single RRAM element. In this figure the black curve shows the predicted current characteristic while the red curve represents the voltage that drops on the bottom electrode. It is evident that initially the whole applied voltage drops at the bottom electrode until $V_{th,1}$ is reached. Then the voltage is divided into approximately 50 % on both elements during the ON-window. After exceeding $V_{th,2}$ the voltage on the bottom element drops to zero, which means that the applied voltage completely drops on the top element. The corresponding reverse effect is observed for negative polarity. Now, the applied voltage drops on the top element until $V_{th,3}$ is exceeded, then the voltage is separated on both elements until $V_{th,4}$ is reached and the voltage drops on the bottom element again.

To integrate CRS devices in a fully passive crossbar array, it needs to be considered how neighboring devices are prevented from unintended switching. Like the effect of sneak paths described in the beginning of this chapter, the applied voltage on a bit line and word line will affect other devices on each line. Thus, it needs to be considered which voltages are applied to the residual bit and word lines. For the case that all other lines are set to ground potential while the addressed bit and word line are set to $+V_{DD}$ and $-V_{DD}$, all devices on these lines experience a voltage drop of $|V_{DD}|$. Only at the addressed device a voltage of $|2V_{DD}|$ drops. Since in this scheme the not addressed devices experience the half voltage of the addressed device, this scheme is called half-select voltage scheme [78]. It is a commonly used method to address devices in cross bar arrays. However, as Figure 2.13b shows that this scheme is potentially not enough for CRS addressing. For very symmetrical RRAM elements in CRS device it holds $V_{th,2} \approx 2V_{th,1}$. Accordingly, the half of a write voltage $V_{\text{write}} > V_{\text{th,2}}$ also exceeds $V_{\text{th,1}}$ and would therefore lie in the ON-window. Thus, a third-select voltage scheme seems to be reasonable for CRS integration. Figure 2.13a shows a cut-out of CRS array that represents the cases of a third-select voltage scheme. The central device is intended to be addressed by a significantly high write voltage V_{write}. That is achieved by applying a write voltage $V_{\text{write}} = V_{\text{DD}}$ to the corresponding bit line and ground potential to the corresponding word line. To ensure that neighboring devices will not be switched, all residual bit lines are set to $\frac{1}{3}V_{\text{DD}}$ and remaining word lines are set to $\frac{2}{3}V_{\text{DD}}$. By this method all not addressed devices experience a voltage drop of $|\frac{1}{3}V_{\text{write}}|$. Like Figure 2.13b shows, this voltage is low enough to prevent the devices from unintended switching to the ON-window.

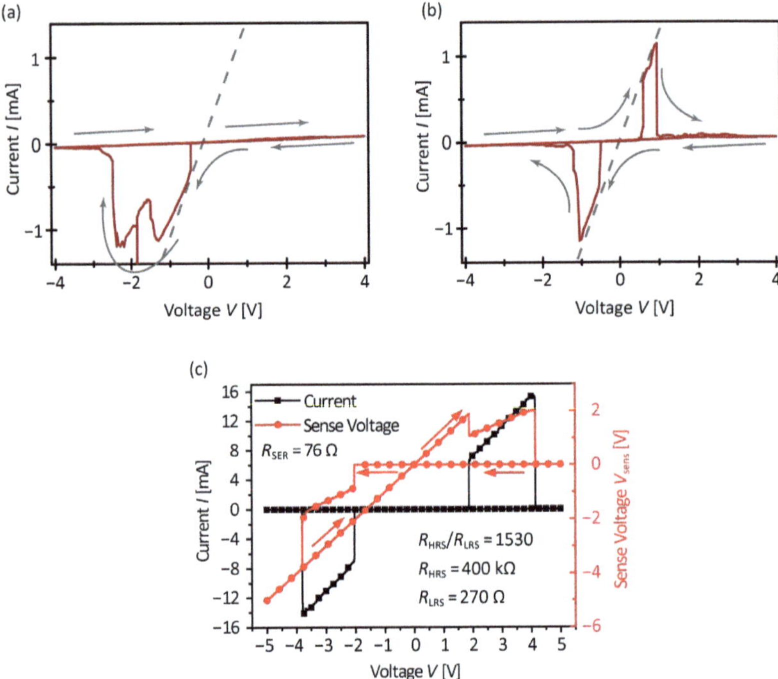

Figure 2.12: Measurement results of integrated CRS cells based on Pt/SiO₂/Cu elements – (a) example of the first *I–V* cycle on a pristine CRS cell and (b) a subsequent curve that shows the typical CRS characteristic. From [75]. (c) *I–V* characteristic of an integrated vertical stack with a series resistor of $R_{\text{ser}} = 76\ \Omega$ to increase the ON-window. As an additional feature a middle electrode monitors the voltage distribution on both single elements. From [73].

However, this method exhibits two major drawbacks. First, it requires a third voltage level that needs to be provided by external circuitry and thus means more effort on the overhead. Second, writing operations need to be performed sequentially on single cross-point junctions. A block wise simultaneous writing operation on a complete word line can only be performed with a half-select voltage scheme.

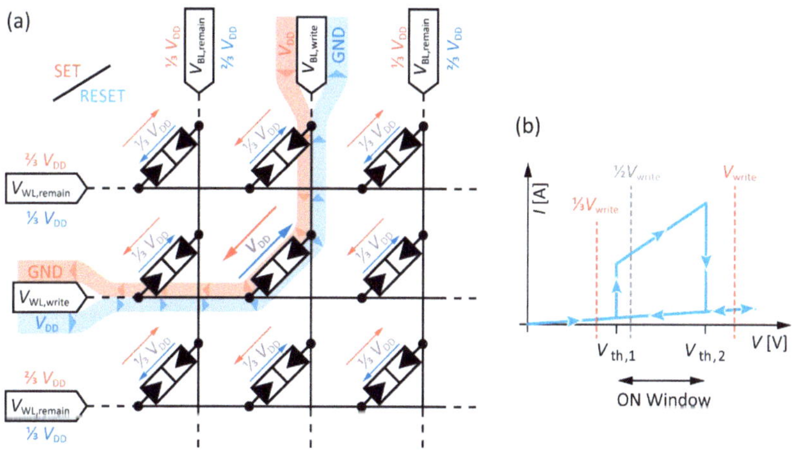

Figure 2.13: Third-select voltage scheme applied on a passive CRS array – (a) In this cut-out of a CRS array the central device is intended to be accessed. With the third-select voltage scheme a significantly high write voltage $V_{DD} = V_{write}$ drops at the central device while all neighboring devices experience the third of this voltage. The red description stands for the SET operation while the blue markings represent the RESET. Arrows at each CRS cross-point junction indicate the polarity for each case. (b) ON-window of the CRS I–V curve with symmetric bipolar resistive switches for positive polarity. While a significant high write voltage V_{write} exceeds the threshold voltage $V_{th,2}$ in order perform a legal CRS operation, the half of its voltage would lie within the ON-window. Unless V_{write} is not too high, the third of its voltage is below the threshold $V_{th,1}$. Adapted from [72].

2.4 Non-Destructive Readout

A drawback of CRS arrays results from the destructive readout which is needed to restore information. Readout is carried out by application of a read voltage which is large enough to switch the cell from state 1 to 0 (identical to write voltage) and is based on subsequent current pulse detection. Statistically, about 50 % of cells read need a write back. Hence,

1.5 write pulses are needed for each read cycle on average which results in high endurance requirements.

2.4.1 General principle

To overcome the drawback of a destructive readout a new CRS setup of two anti-serial resistive switching elements was suggested. In this approach both elements are similar by means of the switching behavior e.g. switching voltages and switching kinetics as well as resistance in the HRS and LRS. However, both elements exhibit different parasitic capacities. A simple equivalent network for such a CRS device is depicted in Figure 2.14.

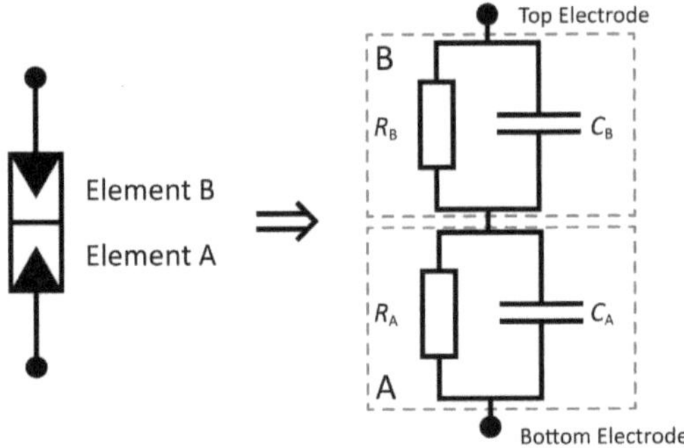

Figure 2.14: Equivalent Circuit of a CRS device – Each element of the CRS device (left) can be described by a resistor (representing the stored resistive state) parallel-connected to a capacitor that is defined by the outer electrodes of the resistive switch. This equivalent circuit is depicted on the right-hand side.

Both resistive switching elements A and B can be described by an element resistance R_A and R_B switchable between the HRS and LRS parallel to an inherent element capacity C_A and C_B, respectively. According to the CRS switching behavior (Chapter 2.3.1) either Element A or Element B may be switched to the LRS or HRS by applying a sufficiently high external voltage between Top Electrode and Bottom Electrode. Therefore, a conductive path is formed or ruptured. Since the thickness of the conductive path is believed to be in the range of a few nanometers [39, 51] its area is practically insignificant compared to the electrode area. Thus, the element capacities remain unaffected.

The capacity of each cell is given by the cell geometry (electrode distance and area) and the material permittivity of the switching material. Variation of the electrode area is certainly the simplest way of fabrication and very useful for experimental proto-type implementation. Practically, varied element capacities may be prepared by different electrode distances and/or the use of materials with different permittivity of each element to decrease the lateral feature size of the CRS device.

In case the resistance of the HRS or LRS state of both elements is similar but the capacities are different, one can distinguish by measuring the time constant of the overall device whether Element A or Element B is in the LRS or HRS without switching or alternatively analyzing the stored information by a capacitive voltage divider.

The basic idea of this concept is outlined in Figure 2.15. An element in the LRS is dominated by the low resistance (corresponding capacitor grayed out) while an element in the HRS is capacitively dominated (high resistive resistor grayed out). The overall capacitance is then determined by the capacitance of the element in HRS. The LRS element is in series to the capacitance. It can be assumed, that the LRS resistance is equal in both elements, so that $R_B('1') = R_A('0')$ holds. If the CRS cell is in state HRS/LRS (cf. Figure 2.15a) the overall capacitance is $C_{cell} = C_A$ while for LRS/HRS (cf. figure 2b) the overall capacitance is $C_{cell} = C_B$. An important difference to other capacitive memories is that information is stored by the value of its capacitance in a nonvolatile way. In other capacitive memory techniques, like DRAM, information is usually stored volatilely by a charge on the capacitance.

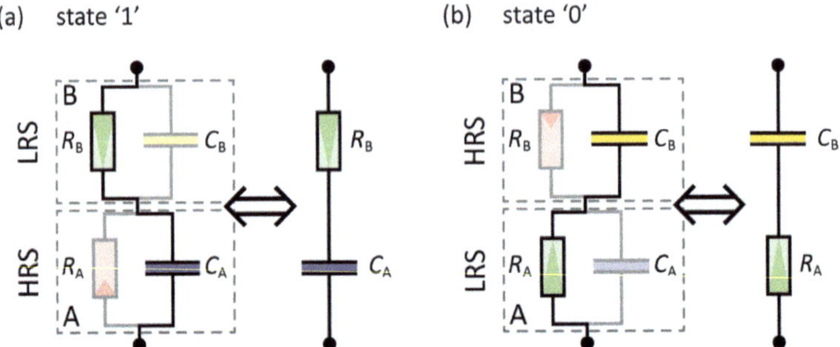

Figure 2.15: Basic concept of non-destructive readout – The equivalent circuit introduced in Figure 2.14 is simplified for the two fundamental CRS states. (a) For state '1', i.e. HRS/LRS, the low resistance of Element B shortens the parallel capacitance C_B. Element A is in the HRS. Thus, the parallel capacitor C_A dominates the overall capacitance. The complete element can be considered as a series circuit of the LRS resistance R_B and the capacitance C_A. (b) Vice versa, CRS state '0' can be considered as R_A in series with C_B.

To read the capacitance C_cell, a voltage pulse is applied to a CRS cell connected to a series capacitor C_out. The cell forms a capacitive voltage divider with Element A (state HRS/LRS) or Element B (state LRS/HRS). Figure 2.16 shows a simple evaluation circuit to measure the capacitance of nanosized CRS implementations.

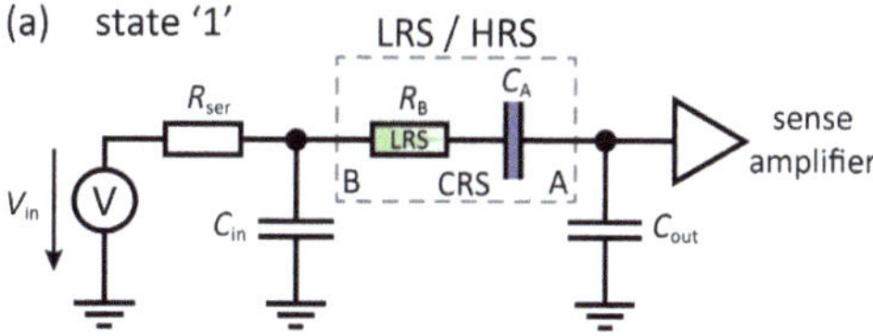

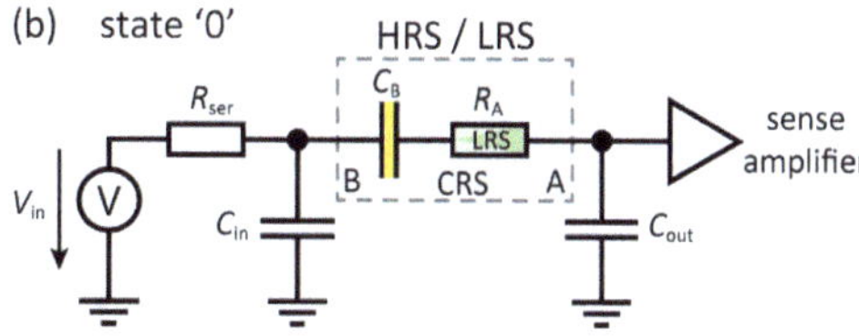

Figure 2.16: Measurement setup for NDRO – the dashed-line box represents the CRS device for both valid states. A signal source, represented by V_{in}, is connected to the CRS device using a series resistor R_{ser} and an input capacitor C_{in}. The sense amplifier is connected to second terminal of the CRS device. The output capacitor C_{out} functions also as reference for the capacitive voltage divider.

The voltage drop at C_{out} is measured by sense amplifier and can be calculated as described in [79]. Involving the simplifications $(R_A \cdot C_A \gg R_B \cdot C_B)$ for state '1' and $(R_A \cdot C_A \ll R_B \cdot C_B)$ for state '0' for high frequency signals ($\omega \to \infty$), the voltage drop at C_{out} is simply

$$\frac{V_{out,0}}{V_{in}} \approx \frac{C_B}{C_B + C_{out}} .$$

(2.1)

The simplifications are motivated by the resistance ratio of R_{HRS} to R_{LRS} compared to the capacitance ratio. Exemplary $R_A = R_{HRS} = 10\ \text{M}\Omega$ and $R_B = R_{LRS} = 130\ \Omega$ which is fulfilled by typical resistance values for SiO_2 [50, 76] as well as $C_A = 0.9\ \text{pF}$ and $C_B = 0.3\ \text{pF}$ based on the device geometry (cf. Figure 2.17).

For the conventional destructive readout of a CRS cell the read voltage is required to be high enough for switching. Typical switching speed of a few nanoseconds is reported in literature [48, 80-82]. However, the nondestructive readout concept is not limited by the switching kinetics and hence fast readout only limited by the RC time constant is conceivable. Moreover, this readout concept is nearly currentless. Therefore, it has the prospect of low power consumption.

2.4.2 Experimental Proof of Concept

The proof which is described in this subsection was demonstrated in [79] is part of the PhD thesis by E. Linn [73].

CRS cells were fabricated using wet-oxidized (SiO_2 thickness 450 nm) p-doped silicon substrates. A 15 nm thick TiO_2 adhesion layer was deposited by DC magnetron sputtering of titanium and subsequent thermal oxidation. The pattern transfer for the 30 nm thick electron beam (e-beam) evaporated Cu electrodes was done using conventional UV lithography and lift-off in acetone, iso-propanol and finally de-ionized water. Afterwards, TiO_2 with a thickness of approximately 8 nm was deposited by high pressure reactive DC sputtering in 74 % Ar and 26 % O_2 atmosphere. Alternatively, 8 nm SiO_2 was deposited by high pressure radio frequency (RF) sputtering using a SiO_2 target in Ar atmosphere. The stoichiometry of TiO_2 and SiO_2 was analyzed by Rutherford Backscattering Spectroscopy (RBS). Finally, 30 nm Pt acting as "top" and "bottom" electrode material was deposited by DC sputtering and pattern transfer was done using UV lithography and lift-off. Using the lateral setup shown in Figure 2.17 two Pt electrodes on the top layer with different electrode areas are combined to a lateral CRS cells by a common Cu "middle" electrode on the bottom layer.

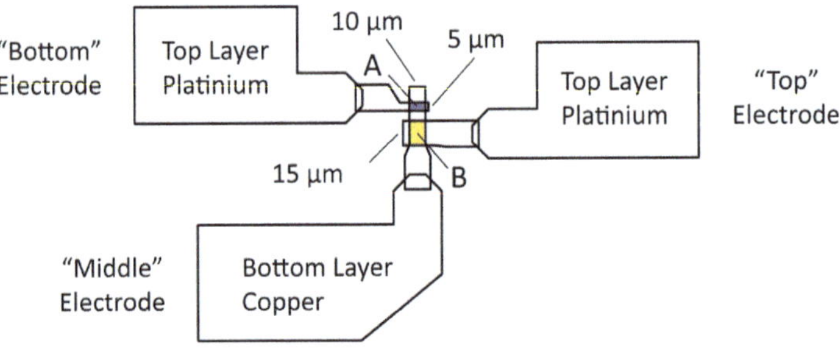

Figure 2.17: Layout of experimental NDRO devices – The CRS device was fabricated in a lateral assembly to reduce process steps during fabrication and to ensure two equal RRAM elements. The single RRAM elements are defined by the intersection of the two top Pt layers with the bottom Cu electrode. The area of the intersecting stripes defines the capacitance of each element. Since the LRS of ECM cells does not depend of the cell area due to the filamentary conductance [83, 84], both resitive switches have nearly same switching properties.

For a quasi-static voltage-sweeps an Agilent Semiconductor Analyzer 4155B was used. The quasi-static voltage-sweeps are only performed to switch between the resistance states and are based on preliminary studies [48, 85]. Due to the Cu electrode the transition between the HRS and LRS state for both SiO_2 and TiO_2 cells is dominated by the diffusion and filament formation of Cu [85]. To proof the new readout approach an Agilent 81110A pulse generator and a Tektronix TDS 684A oscilloscope with active probe head Tektronix P6247 were used. The cable length between the sample and the sense amplifier was minimized to approximately 5 cm. The custom-tailored voltage follower was based on the Texas Instrument OPA 656 operational amplifier with very high input resistance (10^{12} Ω), low differential input capacitance (0.7 pF) and high unity-gain bandwidth (500 MHz). The voltage follower is acting as a sense amplifier which is directly connected to the probe tips and the active probe head of the oscilloscope. C_{out} of the overall sense amplifier circuit includes a discrete ceramic SMD capacitor of 10 pF and was measured using a HP 4284A LCR meter at a frequency of 1 kHz, an AC amplitude of 1 V and 0 V DC bias. C_{out} was found to be 24 pF in this specific measurement setup.

Measurement

The measurement is performed by applying a voltage pulse and the voltage level V_{out} is evaluated. The result is in good agreement to expected values from Eq. (2.1) based on the device geometry and material properties. $Cu/TiO_2/Pt$ based CRS cells are investigated as well as $Cu/SiO_2/Pt$ as switching material in a planar configuration (cf. Figure 2.17) to prove the feasibility of this approach.

The corresponding dielectric constant of TiO_2 in the used cells is approximately 23 which is in the range of values reported for amorphous TiO_2 [86]. Results are depicted in Figure 2.18a (straight lines). From layout geometry and measurements of single resistive switches parameters could be extracted and were used for circuit simulation (dashed line). To verify device behavior not only HRS/LRS and LRS/HRS state were considered, but also the initial HRS/HRS state was measured and simulated with Spice Opus.

The overall capacitance in HRS/HRS can be calculated as follows,

$$C_{cell} = \frac{C_A C_B}{C_A + C_B} \tag{2.2}$$

Since the layout shown in Figure 2.17b was used, the electrode area and correspondingly the capacitance of element A and B should differ by a factor of three, which can be also calculated from measured data:

$$\frac{C_B}{C_A} = \frac{V_{out,0}}{\left(V_{in} - V_{out,0}\right)} \cdot \frac{\left(V_{in} - V_{out,1}\right)}{V_{out,1}} \tag{2.3}$$

From Figure 2.18 we can see that state '0' and '1' are distinguishable by pulse voltage levels.

Measurements with SiO_2 as switching material were conducted as well. Here, the voltage levels due to smaller dielectric constant was lowered. However, the advantage of SiO_2 is that it is a widely used material in semiconductor fabrication. The corresponding dielectric constant of SiO_2 in this study was estimated to be approximately 5.5 which is slightly higher than the literature value for amorphous silica [87]. The measurement was done at ambient conditions hence, incorporation of water could increase the dielectric constant. From Figure 2.18b one can see that state '0' and '1' are distinguishable, nevertheless.

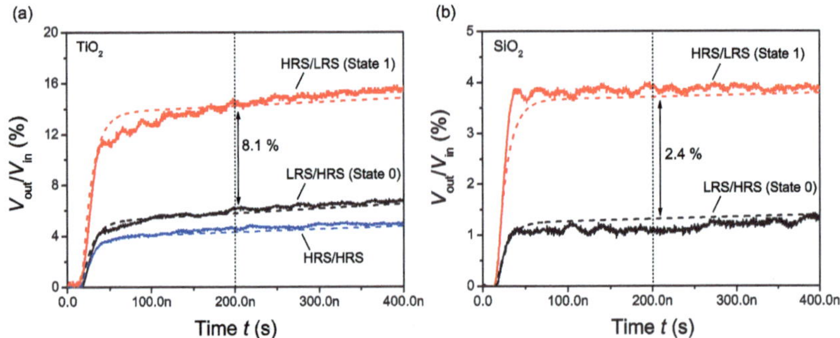

Figure 2.18: Capacitive readout of TiO_2 (a) and SiO_2 (b) CRS cells – The area of Element A is $A_A = 50\ \mu m^2$ and area of Element B is $A_B = 150\ \mu m^2$, film thickness $d \approx 8$ nm, and $C_{out} = 24$ pF. $R_{LRS} \approx 130\ \Omega$, $R_{HRS} \approx 1\ M\Omega$ for TiO_2 and $R_{LRS} \approx 130\ \Omega$, $R_{HRS} \approx 10\ M\Omega$ for SiO_2. In both cases a pulse of $V = 0.5$ V was applied. For TiO_2 (a) $C_A \approx 3.8$ pF was found and $C_A \approx 1.3$ pF which are calculated according to Eq. (2.1). For HRS/HRS $C_{cell} = 0.95$ pF holds. The normalized voltage margin is 8.1 %. For SiO_2 (b) $C_A \approx 0.9$ pF and $C_A \approx 0.3$ pF were found and a normalized voltage margin of 2.4 %.

The voltage level can be improved by use of an optimum capacitor $C_{out,opt}$ which can be found by maximizing the normalized voltage margin

$$\frac{\Delta V_{out}}{V_{in}} = \frac{C_{max}}{C_{max} + C_{out}} - \frac{C_{min}}{C_{min} + C_{out}} \quad \text{with } C_B = C_{max} \text{ and } C_A = C_{min}. \tag{2.4}$$

The maximum of this function is achieved for

$$C_{out,opt} = \sqrt{C_{min} C_{max}}. \tag{2.5}$$

For optimum capacitor $C_{out,opt}$ the maximum achievable normalized voltage margin is

$$\left[\frac{\Delta V_{out}}{V_{in}}\right]_{max} = \frac{1}{1+\sqrt{\dfrac{C_{min}}{C_{max}}}} - \frac{1}{1+\sqrt{\dfrac{C_{max}}{C_{min}}}} \tag{2.6}$$

which is 26.8 % for $C_B/C_A = 3$. Compared 2.4 % for SiO_2 (see figure 3b) and 8.1 % for TiO_2 (see figure 3a) in the first experiments voltage margin can be increased significantly by selection of an optimized C_{out}.

When applying a capacitive readout procedure to a crossbar array, additional capacitances of neighboring cells must be included. A possible readout scheme for a capacitive readout of a crossbar array is shown in Figure 2.19a. Not accessed rows and columns are set to ground, while the accessed row is connected to a serial capacitor and a sense amplifier. All capacities of the accessed column add up to the C_{out} (comprises constant bit line capacitance), while all capacitances in the accessed row add up to C_{in} (comprises constant word line capacitance), hence readout is very similar to reading a single CRS cell described above.

The normalized voltage margin for the worst case of a $m \times m$ array is:

$$\frac{\Delta V_{out,worst\,case}}{V_{in}} = \frac{C_{max}}{m \cdot C_{max} + C_{out}} - \frac{C_{min}}{m \cdot C_{min} + C_{out}} \tag{2.7}$$

The optimum C_{out} then is

$$C_{out,opt} = m \cdot \sqrt{C_{min} C_{max}} \tag{2.8}$$

Hence, the achievable normalized voltage margin for worst case is then

$$\left[\frac{\Delta V_{out}}{V_{in}}\right]_{max} = \frac{1}{m + m \cdot \sqrt{\dfrac{C_{min}}{C_{max}}}} - \frac{1}{m + m \cdot \sqrt{\dfrac{C_{max}}{C_{min}}}} \tag{2.9}$$

In Figure 2.19b the data from TiO_2 cells is taken and an 8×8 array is simulated. C_{out} is 17.5 pF and the voltage normalized margin is 3.3 %. The possible voltages for reading a '1' are between blue straight and dashed line and voltages for reading a '0' are between the black straight and dashed line. For memory arrays with higher element number the voltage margin is decreased significantly. Therefore, this nondestructive readout scheme is only applicable for small to medium sized passive memory arrays.

To counteract this trend, the achievable normalized voltage margin can be increased by increasing C_B/C_A. E.g. a normalized voltage margin of 6.5 % is calculated for $C_B/C_A = 10$

(cf. Figure 2.19c). This capacity ratio is still easily achievable at the same feature size in common semiconductor fabrication using the same material system for both elements but different thickness of the insulator (e.g. 5 nm vs. 50 nm). Even higher capacity ratio may be prepared by further material or device engineering.

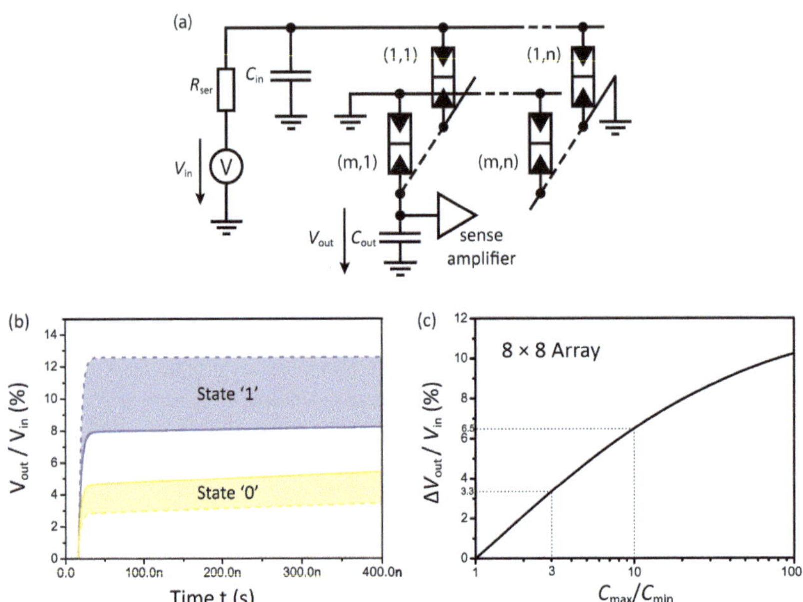

Figure 2.19: (a) Equivalent circuit of crossbar array and capacitive readout scheme. Not accessed lines are grounded, while the accessed row is connected to input voltage V_{in} and the accessed column is connected to series capacitor C_{out}. Output voltage V_{out} is sensed at capacitor C_{out}. It is assumed that $R_{HRS} \gg R_{LRS}$. (b) Simulation of a 8 × 8 array of TiO_2 cells and $C_B/C_A = 3$. C_{out} is 17.5 pF and the voltage margin is 3.3 % according to Eq. (2.9). The worst case is marked by straight lines while best case is marked by dashed lines. Possible levels in between are shaded. (c) Influence of C_B/C_A on normalized voltage margin for an 8 × 8 array. Larger differences in capacitances increase voltage margins. $C_B/C_A = 3$ results in 3.3 % while $C_B/C_A = 10$ results in 6.5 %.

It should be highly emphasized that the preparation of a fast sense amplifier with a desirable small input capacity as well as high input resistance and optimized for a small voltage margin reported above is challenging. Alternatively, the read voltage might be increased using ultra short voltage pulses (with prospect of sub ns-range) for readout resulting in higher voltage levels for state '0' and '1'. This is attributed to the switching kinetic of many resistive switching materials [5, 39, 88] including TiO_2 [67] and SiO_2 [89]. In general, higher switching voltages are needed when decreasing the write pulse length. Hence, it is suggested to apply higher voltages for readout at very short read pulses although these voltages are sufficiently high enough to switch the device at higher write pulse length. The memory cells analyzed in this study for example were switchable at about 1 V using quasi-static voltages-sweeps. However, nondestructive readout cycles applying voltages higher than 1 V with a pulse width of 50 ns without switching the device could be performed. This behavior is illustrated exemplarily for a SiO_2 based asymmetric CRS cell configuration. Figure 2.20a depicts the response V_{out} using read pulse height of $V_{in} = 2$ V and $V_{in} = 3$ V. The response signals are normalized to the individual V_{in}, respectively shown in Figure 2.20b for comparison. It is mentionable that the complete evaluation can be reliably carried out on a sub-µs time scale.

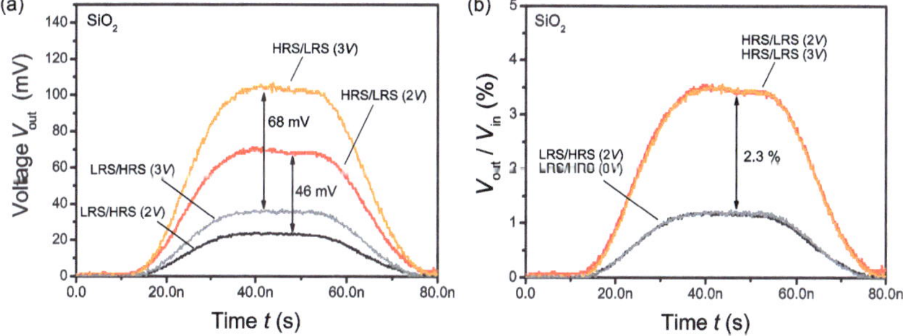

Figure 2.20: (a) Output voltage V_{out} for short pulse lengths (50 ns) applied to SiO_2 cells and pulse heights of $V_{in} = 2$ V and $V_{in} = 3$ V, respectively. Voltage margins are about 46 mV for 2 V input pulse and 68 mV for 3 V input. (b) Normalized output voltages derived from (a) for comparison.

Summary

An approach for non-destructive reading of CRS cells was demonstrated. This has been validated using a setup of two resistive switches with varying cell geometries in an anti-serial assembly. Instead of alternating the cell dimensions to change the capacity, also the material properties of the electrolyte could be varied. Here, the concept of a capacitive readout procedure for single cells was proven by measurements on TiO_2 and on SiO_2 based planar CRS cells. The readout of an 8×8 memory array was proven by simulations derived from the experimental single cell results. This result is an important outcome for the feasibility of the concepts and experiments which are presented in the following chapters of this work.

Furthermore, it was experimentally shown that voltage levels for readout can be increased due to the switching kinetics effects of ReRAM cells for shorter pulse lengths and with higher excitations voltages. In general, this concept has the prospect of a fast (sub-µs) and power-saving non-destructive readout scheme for small to medium sized crossbar arrays without a read cycle limitation due to low endurance.

3   Associative Capacitive Networks

The nature of the capacitive non-destructive readout of CRS cells offers a technique that enables pattern recognition in a weighted synaptic-like manner. Passive associative memories based on weighted interconnects have already been investigated since the late 1980s [14, 90]. Since the technique described in the following is capacitance based, the new developed memory class is called Associative Capacitive Network (ACN).

3.1   Basic Idea

In this concept, a bit-by-bit XNOR operation between input (search pattern) and stored values (templates) is the basic element [91]. The mechanism is illustrated for the core cell in Figure 3.1a featuring two CRS cells for the previously introduced NDRO method. It contains the information template T_{ij} and a redundant negated part $\overline{T}_{ij} = N_{ij}$. The input x_j and its negate $\overline{x}_j$ are applied to the cell. As described later, both parts, T_{ij} and the complement $\overline{T}_{ij}$ (x_j and its complement $\overline{x}_j$, respectively), are essential to ensure that proper matching can be performed in the stored information.

Figure 3.1b shows the general system's architecture. The memory can be assembled in a fully passive CRS crossbar array with the dimensions $2M \times N$, which represents the ACN. Within this area no active CMOS device is used. Controlling is realized in periphery. The search pattern X as well as its complement $\overline{X}$ are applied to the vertically arranged bit lines. The readout takes place on the horizontal connection, i.e. the match line (ML). Depending on the similarity of search pattern and stored information, the match line can reach different voltage levels. That means that this technique internally does not operate with digital signals. Or, in other words, this part of the matching process is performed by analogous signals. However, a suitable CMOS-based Winner-Loser Distance Amplifier (WLDA) together with a following Winner-Take-All (WTA) circuit can digitalize the output instantly [92]. Various implementations are possible and have been shown, mixed digital/analog ones [93-95] as well as fully digital methods [96, 97]. In this work, the idea of a voltage-level to time-domain translation [93] is realized by a simple CMOS comparator (cf. Figure 3.1a). It compares the ML voltage with a linearly rising voltage ramp. In this case a counter triggered by the comparator output ensures the digital unity. This method imposes significant area overhead to satisfy resolution of detection. Counters also

require a high-speed clock. For higher or faster integration this comparator-based implementation can be replaced by more powerful WLDA and WTA circuity.

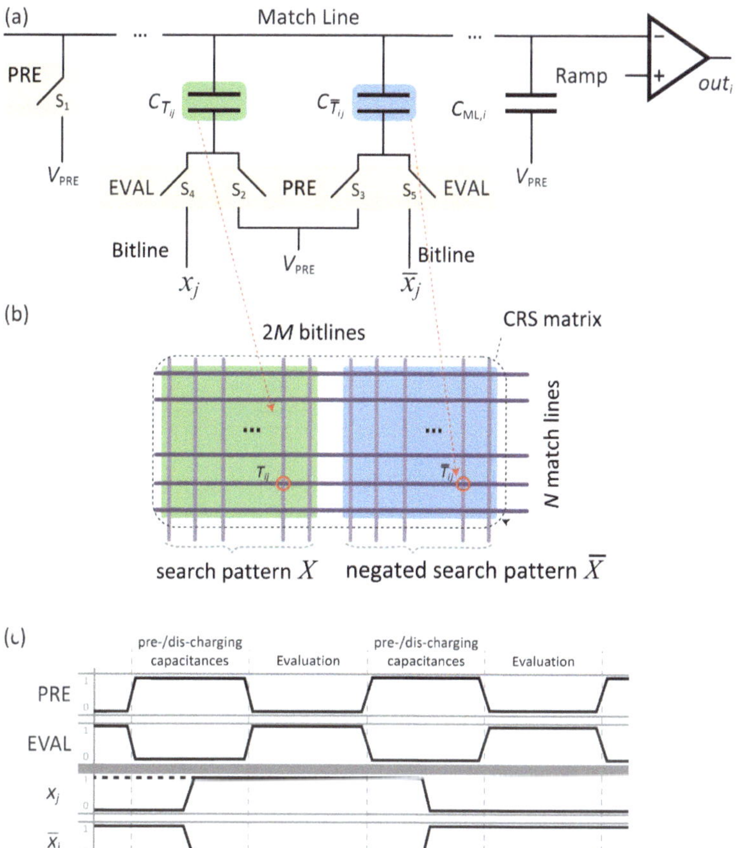

Figure 3.1: General concept of one ACN bit storage – each single information is stored in two CRS devices, here already illustrated by its corresponding capacity. (b) The memory array can be divided into an information (green) and a redundancy (blue) block. For two defined capacity states each redundant capacity must have the negated capacitance compared to the information. For readout (a) the match line is pre-charged by a defined voltage V_{PRE}, here, $V_{PRE} = 0$ V. PRE and EVAL describe the two basic phases. Subsequently, the search pattern is applied to the CRS cells. Controlling takes place by the symbolic switches S_1– S_5. When voltage is applied to the ACN cell, the capacitances $C_{T,ij}$, $C_{\overline{T},ij}$ and $C_{ML,i}$ build a capacitive voltage divider which causes, depending on T_{ij} and x_j, a certain voltage level on the match line (ML). (c) The signal diagram corresponds to the circuit in (a). Pre-charge and evaluation phases are performed subsequently. The match line voltage $V_{ML,i}$ is then related to a rising voltage ramp (cf. Sec. 3.3 and 3.4). Adapted from [91].

The size of a single ACN array is limited due to increasing blurring effects in large passive networks (see also Chapter 3.3). Full integration of large ACN memories requires a fabrication in several blocks. Each block has its own WLDA and WTA stage. This system architecture promises a significant energy efficiency [17, 92], and also allows to increase the total number of inputs by increasing the number of blocks independently of the number of inputs M.

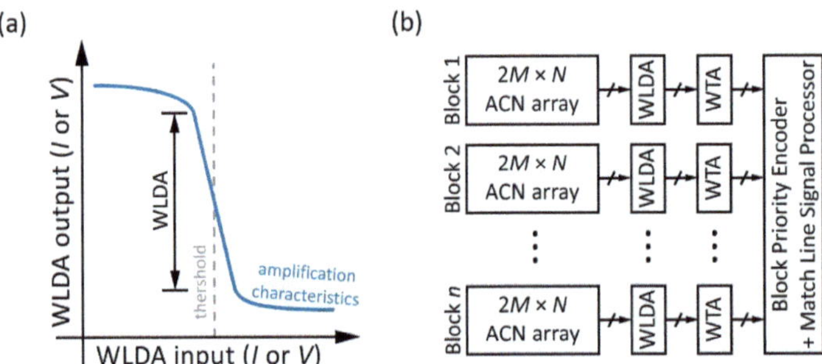

Figure 3.2: (a) Typical characteristic of Winner-Loser Distance Amplification and (b) exemplary circuit of a CMOS based implementation [92].

Unlike the basic functionality of CAMs, ACNs detect the similarity of two binary patterns. This similarity, is called Hamming Distance (HD) in information theory. The Hamming Distance is defined as

$$HD_i = \sum_{j=1}^{M} \left| x_j - T_{ij} \right|.$$

(3.1)

It states how many single bits of two patterns deviate from each other. In this work a test setup for analyzing ACNs that contain all possible Hamming Distances (HD $\in \{0,\dots,M\}$) between X and stored information T.

Concept of Weighting by Capacitances

To demonstrate the capability of finding the nearest match, a concept of weighting is introduced that makes calculations in capacitive networks more understandable. It is related to the concept of synaptic-like calculations, which can be applied to this type of networks, too.

The output voltage of the capacitive network as displayed in Figure 3.3a can be calculated by the principle of superposition of multiple capacitive voltage dividers. The reference voltage V_{SS} can be set to 0 V or ground. The capacitances that are connected between V_{out} and V_{SS} are summarized to C_{SS}. The total capacitance between V_n (which stands for V_1, V_2 and V_3) and V_{SS} is named C_{total}.

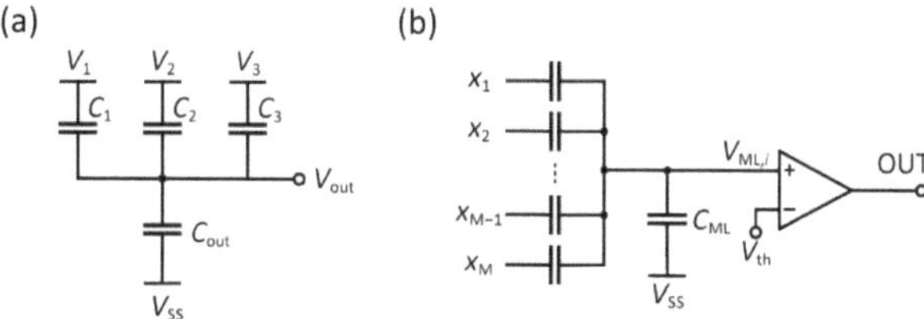

Figure 3.3: Concept of capacitive weight – (a) general principle of a capacitive voltage divider. Appling input voltages to the capacitors C_1–C_M.

So, the total output voltages $V_{out,n}$ can be calculated with

$$V_{out,1} = V_1 \cdot C_{total} \frac{1}{C_{SS}}$$

$$= V_1 \cdot \frac{C_1 \cdot (C_2 + C_3 + C_{out})}{C_1 + C_2 + C_3 + C_{out}} \cdot \frac{1}{C_2 + C_3 + C_{out}} \tag{3.2}$$

$$= V_1 \cdot \frac{C_1}{C_1 + C_2 + C_3 + C_{out}} = V_1 \cdot w_1$$

$$V_{out,2} = V_2 \cdot \frac{C_2}{C_1 + C_2 + C_3 + C_{out}} = V_2 \cdot w_2 \tag{3.3}$$

and

$$V_{out,3} = V_3 \cdot \frac{C_3}{C_1 + C_2 + C_3 + C_{out}} = V_3 \cdot w_3 \tag{3.4}$$

to

$$V_{out} = V_{out,1} + V_{out,2} + V_{out,3} \tag{3.5}$$

As equations (3.2)–(3.4) reveal, the influence of V_n is determined by its input-capacitor C_n. The capacitance ratio in each equation can be summarized in a specific weight w_n, which is dominated by the according input-capacitance. The individual voltage influence of this 3-input example can also be generalized to n inputs:

$$V_{\text{out},n} = V_n \cdot \frac{C_n}{C_{\text{out}} + \sum_{i=1}^{m} C_i} = V_n \cdot w_n \tag{3.6}$$

For capacitive networks with one central node, like in Figure 3.3a, the previously described weight concept can be applied. It simplifies the calculation of the output-voltage V_{out}. It is mandatory that the weight w_n is a dimensionless value and must not by higher than one. Moreover, it holds

$$\sum_{j=1}^{m} w_j = \sum_{j=1}^{m} \frac{C_j}{C_{\text{out}} + \sum_{i=1}^{m} C_i} = \frac{\sum_{i=1}^{m} C_i}{C_{\text{out}} + \sum_{i=1}^{m} C_i} = \frac{1}{\frac{C_{\text{out}}}{\sum_{i=1}^{m} C_i} + 1} < 1 \tag{3.7}$$

and

$$\sum_{j=1}^{m} w_j = 1 \quad \text{for} \left(C_{\text{out}} \ll \sum_{i=1}^{m} C_i \right) \tag{3.8}$$

as in this capacitive network the output voltage V_{out} must be equal or lower to the highest applied voltage V_n (cf. Eq. (3.5) with Eq. (3.2)–(3.4) plugged into Eq. (3.5)).

This concept can be adapted to the idea of an artificial neuron (cf. Figure 1.1 and Figure 3.3b, respectively). The input voltages are weighted by the corresponding input capacitances. Keeping in mind that the capacitances are supposed to be programmable to different values, one difficulty for the calculation of the weight is, that each weight is a function of the residual capacitances. In other words, the individual weight depends on the remaining stored states. As shown later, this behavior must not be necessarily a drawback, but here a simple practice is shown to overcome it.

The applied voltage V_n is either in the logic high state (V_{DD}) or floating, i.e. a high-ohmic input is ensured. Thus, there is no voltage drop between the addressed inputs having the same potential V_{DD}. The unaddressed capacitances do not impact the calculation as well as the corresponding input voltage is floating. Accordingly, the influence of the residual capacitors can be deleted, and derived from Eq. (3.2)–(3.4) it holds:

$$V_{\text{out},1} \left(V_n = V_{\text{DD}} \vee \text{floating} \right) = V_1 \cdot \frac{C_1}{C_1 + C_{\text{out}}} \tag{3.9}$$

$$V_{\text{out},2}\left(V_n = V_{\text{DD}} \vee \text{floating}\right) = V_2 \cdot \frac{C_2}{C_2 + C_{\text{out}}}, \tag{3.10}$$

and

$$V_{\text{out},3}\left(V_n = V_{\text{DD}} \vee \text{floating}\right) = V_3 \cdot \frac{C_3}{C_3 + C_{\text{out}}}. \tag{3.11}$$

Equation (3.6) simplifies to

$$V_{\text{out},n} = V_n \cdot \frac{C_n}{C_{\text{out}} + C_n} = V_n \cdot w_n. \tag{3.12}$$

This procedure is utilized in the experimental part in Section 3.3 and 3.4, since the investigation with less vulnerability to failure is beneficial. Otherwise, each single device defect would cause a failure of the whole test array. In general, various voltage schemes are conceivable, as long as each match line is suitably pre-charged.

Applied to a generalized array structure, each word line can be represented by Figure 3.3b. For the considered search application, the central node in the voltage divider is named match line (ML). Accordingly, C_{out} corresponds to C_{ML}.

3.1.1 CRS-based Concept

The implemented CRS-based ACN is operated in two basic phases, a pre-charge phase (PRE) and an evaluation phase (EVAL). For the first phase PRE, the whole array is set to a defined voltage level V_{PRE}, which is for ease of use set to ground potential. This phase ensures that the initial conditions of each search iteration are the same. During EVAL, the actual search pattern X and its negate $\bar{X}$ are applied to the ACN. Figure 3.1a,c illustrates these phases. The phases are realized by the switches S_1– S_5, which can be implemented by customized CMOS circuits. Switches S_2– S_5 are connected to the bit lines, whereas switch S_1 is connected to the match line. As bit lines and match lines are shared for all cells of an array, in total $\#S = 4M + N$ switches are needed. As described earlier, each match line is fed into a comparator that can be realized by an operational amplifier circuit. By exceeding the preset threshold voltage V_{th}, the output is pulled up to the positive operation voltage. If required, a small hysteresis realized by a Schmitt trigger gate can be implemented and suppress wobbling effects. It is convenient to take a linear rising voltage ramp as the threshold voltage. By that, different voltage levels on all match lines can be found. The specifications of this threshold signal can be found out with a characterization process for a specific size M of the search word.

The two CRS cells represent one stored bit and its complementary. As discussed earlier, the logic state of a CRS cell defines the cell capacitance. This capacitance is connected to input x_j and match line ML_i. Here, $C_{T,ij}$ and $C_{\bar{T},ij}$ represent the effective capacitances of one bit of a stored template T and its negate $\bar{T}$, respectively. These capacitors shape a capacitive coupling between columns and rows, and each one of them acts as a synaptic connection between input pulse and output neuron (cf. Chapter 1 and Figure 1.1) [91]. During the PRE-phase both input signals x_j and $\bar{x}_j$ are decoupled from the structure. During the EVAL period either x_j carries a logical '1' and $\bar{x}_j$ a '0' or vice versa. In the second phase, the evaluation is carried out. Now, the synaptic contributions of the network accumulate and increase the voltage of the match line ML_i.

The voltage $\Delta V_{ML,i}$ on the match line for this phase can be expressed as

$$\Delta V_{\mathrm{ML},i} = \Delta V_x \cdot \frac{\sum_{j=1}^{M} x_j C_{T_{ij}} + \bar{x}_j C_{\bar{T}_{ij}}}{\sum C_i} \tag{3.13}$$

where x_j and $\bar{x}_j$ are 0 for the phase PRE. x_j and $\bar{x}_j$ have a complementary character but can be defined for various values. For ease of use, it is reasonable to define them by the set $\{0,1\}$. ΔV_x is the value of the voltage pulse applied to the bit lines. The capacitance $\sum C_i$ is the capacitance of the whole match line and can be calculated by

$$\sum C_i = C_{\mathrm{ML},i} \cdot \sum_{j=1}^{m} \left(C_{T_{ij}} + C_{\bar{T}_{ij}} \right) \tag{3.14}$$

where $C_{\mathrm{ML},i}$ is the match line capacitance.

The match line voltage is fed into the positive input of a comparator for each match line. The output *out* is activated whenever the sum of the synaptic inputs exceeds the threshold voltage V_{th}, so that for *out* holds:

$$out_i = \begin{cases} V_+ \triangleq \text{logic } 1, & \text{if } \Delta V_{\mathrm{ML},i} > V_{\mathrm{th}} \\ V_- \triangleq \text{logic } 0, & \text{if } \Delta V_{\mathrm{ML},i} \leq V_{\mathrm{th}} \end{cases} \tag{3.15}$$

This is a one-to-one implementation of a simple artificial neuron, as shown in Figure 1.1, where $C_{T,ij}$ and $C_{\bar{T},ij}$ are synaptic weights and EVAL is an activation function with an external threshold, offset or bias [98]. In general, the activation function can be a linear, a step, a ramp, a Gaussian or a sigmoid function. The output circuitry for spiking networks and for performing cognitive tasks can be designed as an Integrate-and-Fire neuron [91].

The variation of the match line voltage corresponds to the Hamming Distance between X and the sorted information T. The Hamming Distance is a practical measure for hardware pattern recognition based on binary and ternary content addressable memories (CAM) (cf. Section 3.2).

3.1.2 2-Bit Example

With the understanding of weighted signals by capacitors, the basic functionality of CRS-based ACNs can be explained by the following 2-bit example. Related to the conventions in Figure 3.1 all four permutations of X are discussed, i.e. $X =$ '00', $X =$ '01', $X =$ '10', and $X =$ '00'. As mentioned earlier, each capacitance stands for a certain information state and has a corresponding weight within the capacitive network. These patterns are stored in the informational part of the memory (cf. Figure 3.4), highlighted by the green box. The redundant negation is stored in the part of the array within the light blue box. The four permutations are illustrated in Figure 3.4b–d. In this example the following equation holds

$$\Delta V_{\text{ML},l} = \frac{C_A}{\sum_i C_i}\left(x_1 + x_2 + c_r\left(\overline{x}_1 + \overline{x}_2\right)\right) \qquad (3.16)$$

the ratio $c_r = C_B/C_A$ is a crucial characteristic for an ACN. It influences the variation of $\Delta V_{\text{ML},i}$ and limits the size of the ACN matrices. If c_r increases, the probability to detect a certain HD increases accordingly (cf. Section 3.3.6).

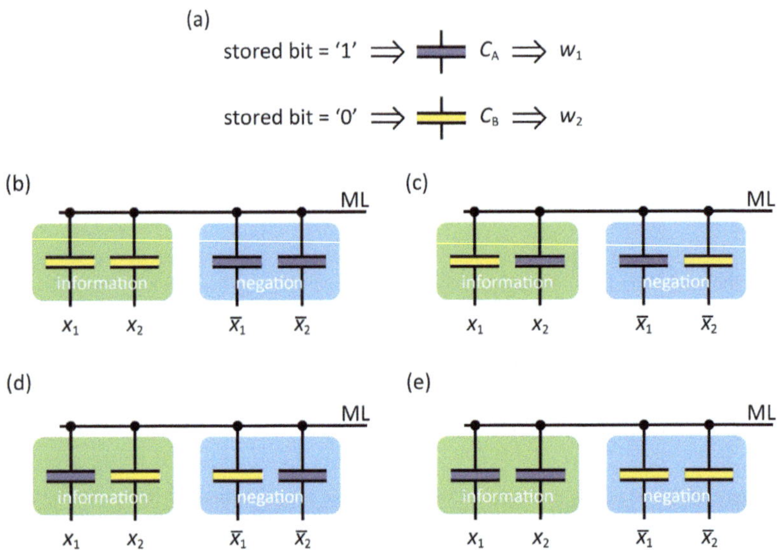

Figure 3.4: Basic concept example – (a) a two-bit ACN example (b–e) the redundant negated information is necessary, as this example explains. Adapted from [91].

Redundant Information

The following example explains why ACNs require a redundant information and how HD detection works. Assuming there is no redundant and negated information stored in the memory, leads to the fact that each single memory cell stores one bit of information. In the case of the here adapted capacitive readout, each stored information bit corresponds to a specific capacitance resulting in a specific weight to the match line. In this example, applied signals can be high (H or '1') or low (L or '0'), representing a certain positive voltage or no voltage, respectively. In general, various schemes for this readout are conceivable, but here the most common standard CMOS-like scheme was chosen. In general, the input x_n can only take two values:

$$x_n \in \{s_1, s_2\}, \quad \text{with} \quad s_1 \neq s_2 \tag{3.17}$$

The same applies to the capacitive weight. Here, the symbolic unit T_n is introduced which stands for the template characteristic of each capacitor. It can take a certain weight:

$$T_n \in \{w_1, w_2\}, \quad \text{with} \quad w_1 \neq w_2 \tag{3.18}$$

In this example, it is assumed that $w_1 = 0.4$ and $w_2 = 0.1$ and the 2-bit array stores the pattern '10', i.e. input x_1 is weighted with $w_1 = 0.4$ and x_2 with $w_2 = 0.1$. As depicted in Figure 3.5 there is no correlation between the summed signals on the match line and the similarity between input signal and stored pattern, i.e. the Hamming distance HD. It could be argued, that for the output $out = 0.4$ it is HD $= 0$ and the deviation in both directions indicates increasing HD. But as shown in Figure 3.5b and d, this is not true. This behavior can also not be adjusted by the balance of weights or alternative bipolar voltage schemes. The conceptual problem is that weights created by capacitances cannot be subtractive but only additive.

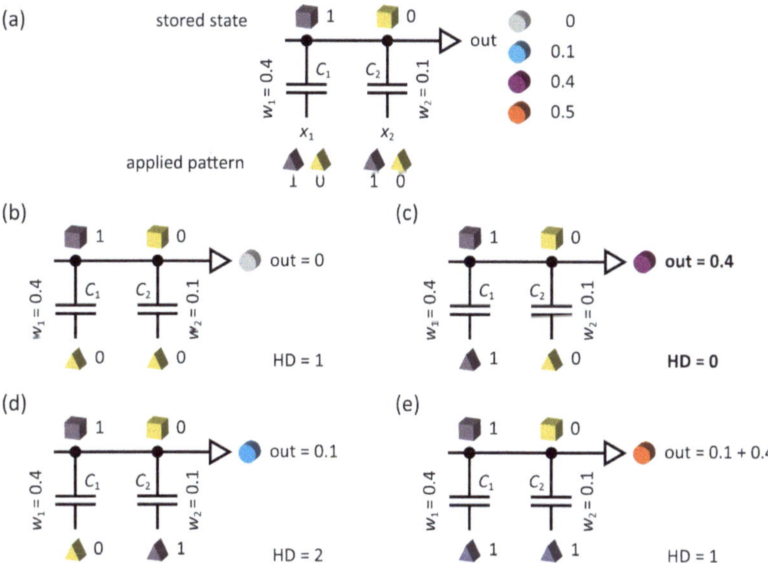

Figure 3.5: Example for capacitive storage without redundancy   (a) the information is stored in single NDRO CRS devices. In this case, the first device stores a logic '1' (here, $w_1 = 0.4$) and the second stores a '0' ($w_2 = 0.1$). The applied signals can either be high (the match line is weighted with the stored information) or low, which corresponds to no weighting. For weights are summarized on the match line. The input combinations '00', '01', '10' and '11' (b-e) cause four different match line outputs.

This observation can be expressed by the following consideration. Each single cell (single capacitor) has four possible combinations of output. With the consideration of Eq. (3.17), the output equation for each single bit is:

$$out_{\text{1bit}} = x_1 c_1 = \left(s_1 \vee s_2 \right)\left(w_1 \vee w_2 \right)$$
$$= s_1 w_1 \vee s_1 w_2 \vee s_2 w_1 \vee s_2 w_2 \, .$$

(3.19)

As inputs and weights must be differentiable by a fundamental rule of logic concepts, the four combinations lead to at least three different outputs, in general even four different outputs. That gives inconclusive information in terms of Hamming Distance. Accordingly, one needs to find a way to work around this problem.

To overcome this issue, the integration of redundant information in the form of a fully negated pattern is exploited. Now, each bit has a negated duplicate. Of course, the readout pattern needs to be extended in the same way as well. In consequence, it is ensured that one part of this "double bit" is always applied by the high signal of the readout pattern. By this, each "double bit" can give a high or low weight depending on the single bit HD. That leads to the output equation:

$$out_{\text{d-bit},n} = x_1 T_1 + x_2 T_{2'}$$
$$= x_1 T_1 + \overline{x_1 T_1}$$
$$= \left(s_1 w_1 + s_2 w_2 \right) \vee \left(s_1 w_2 + s_2 w_1 \right) \vee \left(s_2 w_1 + s_1 w_2 \right) \vee \left(s_2 w_2 + s_1 w_1 \right)$$
$$= \left(s_1 w_1 + s_2 w_2 \right) \vee \left(s_1 w_2 + s_2 w_1 \right)$$

(3.20)

This fuzzy logic function (cf. Chapter 4) corresponds to the basic XNOR function of conventional logic. Figure 3.6 shows the conventional symbol, the truth table and the mathematical function of a XNOR-gate.

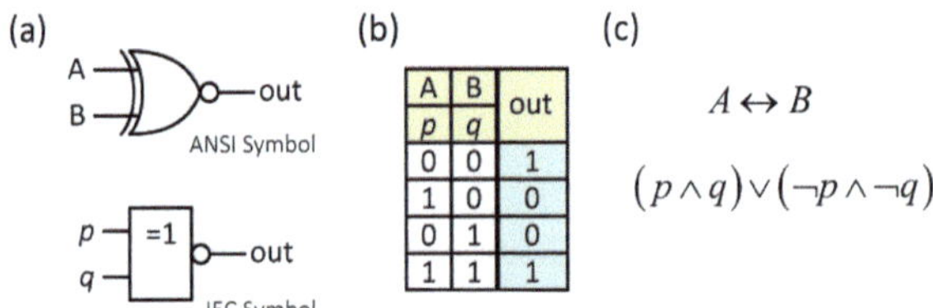

Figure 3.6: classical XNOR-gate (a) symbol (b) truth table and (c) mathematical function (adapted from [99].

Each "double bit" can, thus, only contribute with two different results. It can be defined as a positive match (i.e. HD = 0) or a mismatch (i.e. HD = 1). This leads to distinct definite results on the match line. An example for this redundant approach is illustrated in Figure 3.7. The bright blue box marks the redundant information part. For better comparability, the same basic weights are chosen as in Figure 3.5, but now the structure of two "double bits" generates only three different results on the match line. The cases shown in Figure 3.7b and Figure 3.7e lead to the same result. As long as the fundamental requirement of negation in the redundant element is fulfilled, the permutation of stored information leads to distinctive results again. The remaining cases can be easily derived by the example in Figure 3.7. This example explains why adding a redundant negated information is essential for the principle of ACNs.

For the sake of completeness, the mathematical description of two "double bits" is carried out. In order not to build all possible permutations, it is convenient to build the sum of two "double bits". With Eq. (3.20) the equation can be simplified to:

$$
\begin{aligned}
out_{2\text{ d-bit}} &= out_{\text{d-bit},1} + out_{\text{d-bit},2} \\
&= x_1 T_1 + \overline{x_1 T_1} + x_2 T_2 + \overline{x_2 T_2} \\
&= \left[\left(s_1 w_1 + s_2 w_2\right) \vee \left(s_1 w_2 + s_2 w_1\right)\right] + \left[\left(s_1 w_1 + s_2 w_2\right) \vee \left(s_1 w_2 + s_2 w_1\right)\right] \\
&= \left(s_1 w_1 + s_2 w_2 + s_1 w_1 + s_2 w_2\right) \vee \left(s_1 w_1 + s_2 w_2 + s_1 w_2 + s_2 w_1\right) \vee \\
&\quad\ \left(s_1 w_2 + s_2 w_1 + s_1 w_1 + s_2 w_2\right) \vee \left(s_1 w_2 + s_2 w_1 + s_1 w_2 + s_2 w_1\right) \\
&= 2\left(s_1 w_1 + s_2 w_2\right) \vee \left(s_1 w_1 + s_2 w_2 + s_1 w_2 + s_2 w_1\right) \vee 2\left(s_1 w_2 + s_2 w_1\right)
\end{aligned}
\tag{3.21}
$$

Obviously, only three outputs can be derived for any two "double bit" configuration. The general equation for n bit structures is:

$$
\begin{aligned}
out_{n\text{ d-bit}} &= \sum_n out_{\text{d-bit},n} = \sum_n \left[x_n T_n + \overline{x_n T_n} \right] \\
&= \sum_n \left[\left(s_1 w_1 + s_2 w_2\right) \vee \left(s_1 w_2 + s_2 w_1\right) \right]
\end{aligned}
\tag{3.22}
$$

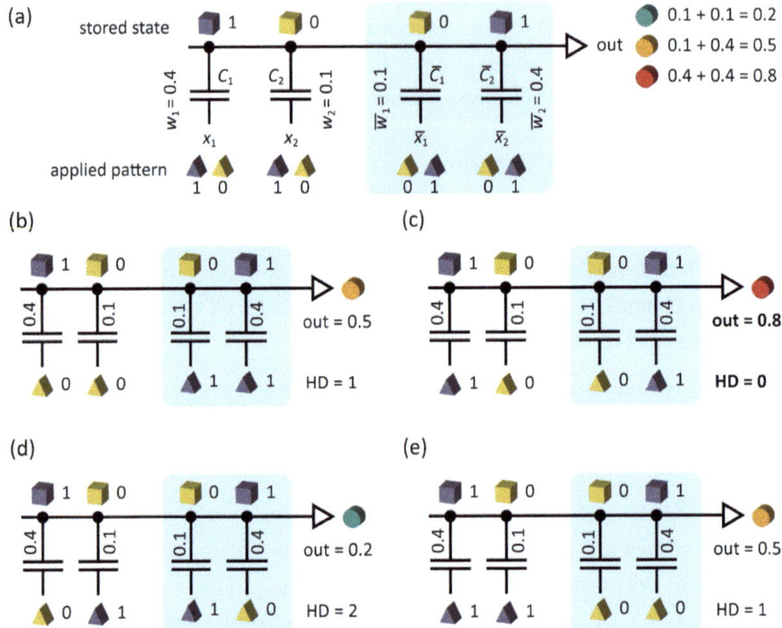

Figure 3.7: Example for capacitive storage with redundant part– (a) in analogy to Figure 3.5 the information is stored in single NDRO CRS devices. The left part of this two-bit cell represents the informational part, the right section contains the negate redundancy. Like in Figure 3.5, the first device stores a logic '1' (here, $w_1 = 0.4$) and the second stores a '0' ($w_2 = 0.1$). Additionally, the third device stores a logic '0' (i.e. the inverse of the first device) and the fourth device stores a logic '1'. The applied signals correspond to the scheme as in Figure 3.5 introduced. Now, the input combinations '00', '01', '10' and '11' (b-e) cause three different match line voltages which correspond to the Hamming Distance between stored and applied pattern.

Circuit Considerations

To realize such a network, it is important to consider some electrical aspects.

Signals are applied with a high impedance output so that no back propagation into the source takes place. That is important for voltage summation on the match line. Otherwise the summation on the match line is not only referenced to the match line capacitor C_{ML} but also to the signal source's output impedance.

For the signal detection, similar conditions are required. A low impedance input would deteriorate the capacitive voltage divider property.

Input and output capacitances of detectors and voltage drivers play an important part in terms of RC-time and accuracy. Variations between individual circuit elements can decisively influence the behavior of the network. Here, a reasonable tradeoff needs to be found.

3.1.3 Different Readout Voltages

Previously, it was assumed that $x_j = $ '1' corresponds to a positive voltage pulse, and $\bar{x}_j = $ '0' to zero voltage. However, different schemes can be adapted in ACNs. Thereby, some general strategies need to be considered. In Table 3.1, different readout schemes are summarized. Assignments of logic states to corresponding capacitances (T) as well as the assignment for the input pattern (X) are iterated. It essentially influences the order in which the nearest matches can be found. Here, the network is always discharged in the PRE-phase. Also, only positive voltages are considered. An extensive list of all thinkable readout schemes can be found in Appendix 1.

Case	PRE	Stored states	Applied pattern	Match line	Time for rising voltage ramp
A	GND	'1' = C_{low} '0' = C_{high}	'1' = V_{DD} '0' = GND	low V = low HD high V = high HD	Match = 1st result
B	GND	'1' = C_{high} '0' = C_{low}	'1' = V_{DD} '0' = GND	low V = high HD high V = low HD	Match = last result
C	GND	'1' = C_{low} '0' = C_{high}	'1' = GND '0' = V_{DD}	low V = high HD high V = low HD	Match = last result
D	GND	'1' = C_{high} '0' = C_{low}	'1' = GND '0' = V_{DD}	low V = low HD high V = high HD	Match = 1st result

Table 3.1: ACN voltage schemes – the definition of logic states in the network and the applied pattern can be defined for individual demands.

3.2   ACNs are Content Addressable Memories

To classify the associative property of ACNs, Content Addressable Memories (CAM) are introduced and the state-of-the-art is described in this section.

A CAM is a special hardware type of memory which differs from other conventional mass storage types of memory, like Random Access Memory (RAM), in its function of data addressing. In this type of memory, the stored information is compared to a certain set of input data. As a result, the output is the address of the matching data. That makes it very effective for very-high-speed searching applications. The general structure of CAMs is illustrated in Figure 3.8.

This Section introduces the basic idea of Content Addressable Memories to reveal the analogy of ACNs to CAMs. ACNs can be utilized to implement a CAM without the use of CMOS-technology in the memory. The basic requirements of a search iteration can be applied to the same block diagram that is valid for CMOS-based CAMs (cf. Figure 3.8). Furthermore, by redoubling every ACN core cell, i.e. the combination of two CRS with opposite logic states (cf. Figure 3.1a), also Ternary CAM arrays (TCAM) are achievable. However, achieving the same information capacity requires twice the area of a standard ACN.

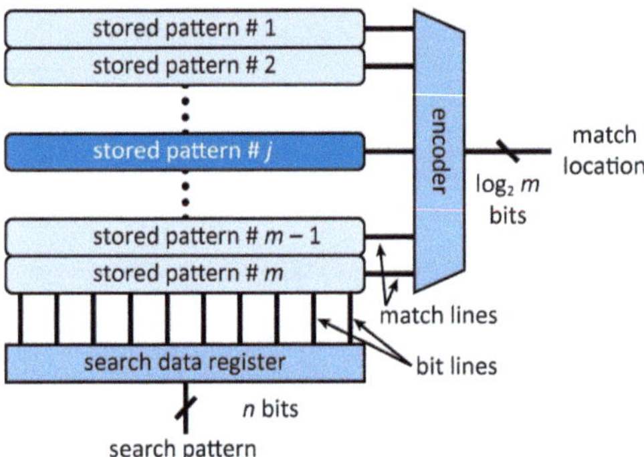

Figure 3.8: General concept of CAMs – A content addressable memory of the dimensions $m \times n$ stores m patterns, i.e. words, each with n bits. The search pattern is applied to the memory via the search data register. All n bits of the search pattern are applied simultaneously to the entire array. In this example the stored pattern # j matches with the search word. The answer of the encoder represents location j of the horizontal match lines. This address is encoded with $\log_2 m$ bits. Adapted from [15]

3.2.1 State of the Art

Content addressable memories have become important in the field of network routing. Until now, the conventional realization is implemented by state-of-the-art CMOS processes. The core cell is based on well-established and widely studied static random-access memories (SRAM). SRAM is eminent for its fast switching speeds and can be easily integrated, which is why SRAM is first choice CPU memory [4]. The major drawbacks are high area demand ($\sim$140 F^2), static current losses and volatility.

Standard CMOS SRAM cells are built of six transistors – two p-channel and four n-channel MOSFETs. The key elements to store information are two cross-coupled inverters. This circuit is also called bistable latching circuitry (flip-flop) [100] and has two stable states, which designates the binary information. Its two nodes need to have complementary potentials – Q and $\bar{Q}$ (cf. Figure 3.9a) – since the output of the one inverter is the input of the other. Since CMOS gates are self-enhancing, the potentials of the nodes are stable – high/low and low/high.

To assemble these memory cells in a random-access memory, two additional pass transistors are needed. These MOSFETs can be controlled by the word line (WL) voltage. A high signal allows the readout of Q and $\bar{Q}$ on two bit lines (BL and $\overline{\text{BL}}$). Common practice is pre-charging both bit lines to a high level. By selecting the word line, one of both bit lines needs to be discharged over the n-channel MOSFETs and the line is pulled low. This differential signal is detected by sense amplifiers which are connected to all bit lines.

To write data into the SRAM, the bit lines are set to the corresponding level, while the condition of opposite voltages is retained. By activating the word line, the stored states are overwritten by the bit line voltages. This can only be achieved when the inverter transistors are comparable weak, which is the case for high integrated CMOS circuitry.

In the setup of SRAM-based CAMs, the previously described 6-transistor memory cell is extended by adding further circuit elements. Basically, it is distinguished between NOR-type (cf. Figure 3.9b) and NAND-type CAMs (cf. Figure 3.9c).

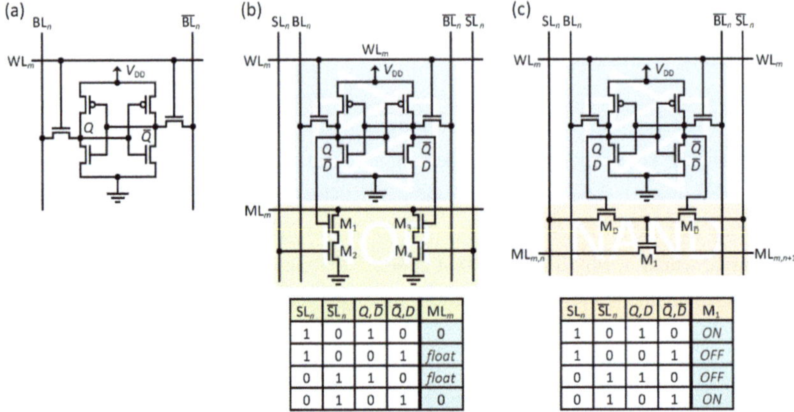

Figure 3.9: SRAM based CAM cells – (a) 6-transisor SRAM cell, the inner two cross-coupled CMOS-inverters have two stable states, indicated by Q and $\bar{Q}$. The two outside transistors serve as access switches, (b) 10-transistor NOR-type CAM cell, the SRAM cell is extended by four transistors that connect the match line ML_m with ground if the search bit matches with the stored bit (c.f. middle truth table) (c) 9-transistor NAND-type CAM cell, the SRAM cell is extended by three transistors which connect the match line segments $ML_{m,n}$ and $ML_{m,n+1}$ if the searched bit matches with the stored bit (c.f. right truth table).

Here, the NOR-type CAM is described in more detail. The same can be applied for the NAND-type CAM.

Like illustrated in Figure 3.9b, four n-channel MOSFETs are assembled like the n-channel part of a classical dynamical CMOS XNOR-gate with the inputs SL and D. The nodes Q and $\bar{Q}$ of the unmodified SRAM cell are connected to the gate of the transistors M_1 and M_3. The transistors M_2 and M_4 are connected to two additional vertical search lines (SL and $\overline{SL}$). The drain of M_1 and M_3 is connected to a further horizontal match line (ML). In principle the order of M_1 and M_2, respectively M_3 and M_4, has no impact on the functionality. To perform a content related search operation, the match line is pre-charged to a high voltage and the searched bit couple is applied to the search lines. Like for the write operation of the SRAM cell, the search bit couple must fulfill the requirement of complementary voltage levels. Now, two general cases can be considered:

$$SL = Q \quad \text{and} \quad \overline{SL} = \bar{Q} \tag{3.23}$$

$$SL = \bar{Q} \quad \text{and} \quad \overline{SL} = Q \tag{3.24}$$

According to (3.23), either SL and Q are high or $\overline{SL}$ and $\overline{Q}$ are high. Consequently, M_1 and M_2, or M_3 and M_4 are activated. In this case the pre-charged match line is pulled to a low potential and is discharged. For (3.24) there is always a closed transistor in the path between ML and ground. Either, M_1 and M_3 are open or M_2 and M_4 interrupt the connection.

In analogy, the circuit of the NAND-type CAM (cf. Figure 3.9c) causes that the match line transistor M_1 is switched on for each match condition. When the condition in Eq. (3.23) is fulfilled, either transistor M_D or $M_{\overline{D}}$ is switched on and thus the gate of M_1 is set to high potential. By this, the match line sections $ML_{m,n}$ and $ML_{m,n+1}$ are joined. When this match condition is met for all CAM cells in one row, the entire match line builds one conductive path. Elsewise, at least one transistor interrupts the nMOS chain.

To perform a content related read operation, the general principle of CAMs needs to be considered. Like demonstrated in Figure 3.8 the search is performed simultaneously on the entire memory array, i.e. the search pattern is applied via the search data register to all search lines and so to all stored patterns. So, when the search register output is activated, all search lines are set to the corresponding levels of the search pattern. In the memory array, all cells of one horizontal row are connected to one match line. Figure 3.10 shows this situation for the NOR-type CAM. All match lines are pre-charged only once during one search cycle. Figure 3.11 gives an example of the applied voltage signals to control a NOR-base CAM. When the search pattern is applied, the transistors M_2 or M_4 of each cell are activated. If the condition in Eq. (3.24) holds for all cells of one stored pattern, the match line remains disconnected from the ground voltage level and the charge on the match line is unaffected apart from minor leakage current. However, if only one single cell fulfills the condition in Eq. (3.23), the whole match line is connected to ground potential and the voltage is pulled down. Discharging the match lines is equivalent with a mismatch, at least for one bit of the search pattern and the stored pattern. When the voltage of the match line can be retained, all bits of the search pattern match with the bits of the corresponding stored pattern. Thus, the result is equivalent with a complete match.

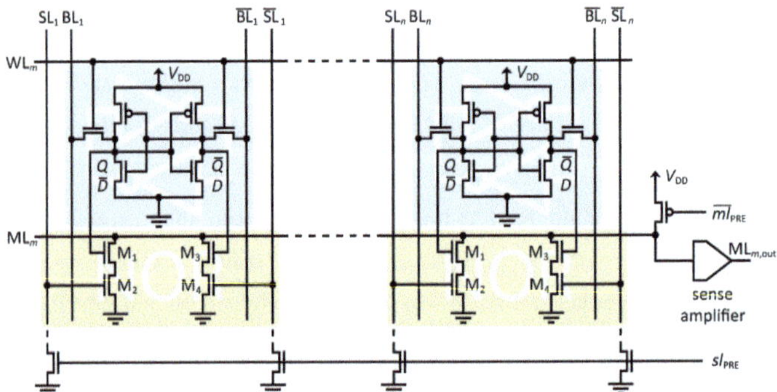

Figure 3.10: Segment of a NOR-type CAM – in each row the word lines (WL) and match lines (ML) of the corresponding pattern are connected. Additionally, $2n$ pulldown transistors allow a pre-charge operation of all search lines (SL) via the signal sl_{PRE}. The match lines are pre-charged by m pullup transistors controlled by the signal $\overline{ml}_{PRE}$.

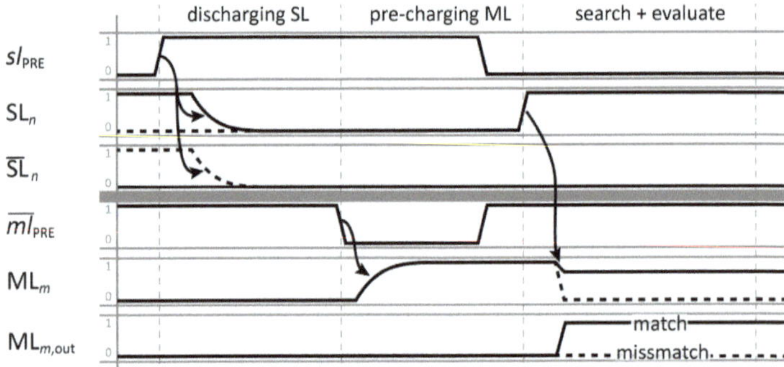

Figure 3.11: Typical voltage signals for NOR-type CAMs – labels and phases match to the circuit depicted in Figure 3.10. The discharging phase of the search lines is controlled by sl_{PRE} and the pre-charging phase of the match lines is controlled by $\overline{ml}_{PRE}$. It must be ensured that both phases are sufficiently long to fully reach the intended conditions. The actual search is performed during the last phase, when the search pattern is applied to the search lines. During this phase SL must be the complement of SL. Depending on the stored information, the match line voltage is retained or pulled to ground voltage level. $ML_{m,out}$ is the rectified voltage by the match line sense amplifier.

3.2.2 Ternary CAM

Ternary Content Addressable Memories (TCAM) extend the functionality of conventional CAMs by adding a third logic state. In practice this logic state is utilized as wildcard for irrelevant information. Therefore, this third logic state is also defined as the "don't care"-bit. This feature is relevant in network switches where certain network ranges are addressed. The search pattern can be reduced to the demanded information part.
The "don't care"-bit can be set in the stored data as well as in the search pattern. That means that either the respectively stored pattern part always returns a match, regardless of the applied search string, or the respective part of the search pattern does not influence the matching result. So, the ternary property needs to be applicable in the storage as well as in the search operation. The truth table for all states and results is shown in Figure 3.12a. A "don't care" is realized by a pair of '0' in search operations. "don't cares" in the storage are realized by two logic '1'.

As the two available states of a flip-flop always represent the complement of each other, TCAMs require an extension to the core cell. In common CMOS-based TCAMs that is achieved by redoubling the flip-flop. The already introduced CAM cell structures still represent the base of the core cell. However, the inputs of the transistors of the logic gate part are now connected differently. For the NOR-type TCAM the only difference is, that M_4 is now connected to the $\bar{Q}$ node of the second flip-flop. Furthermore, additional vertical control lines are required to write information into the second flip-flop. Depending on the architecture, few transistors can be saved compared to a full duplication. The "don't care" is generated when both flip-flops store the same information or when both search lines, SL and SL, have the same logic state. Figure 3.12a shows the truth table that is valid for all TCAMs. Besides the operations that were already shown for CAMs, two additional cases are introduced. In principle, redoubling the flip-flop adds more cases than shown the system, however, only one additional state is required. Thus, all remaining constellations are not allowed. As convention for CMOS-based TCAMs, the "don't care" in the storage is represented by $Q = \bar{Q} = $ '1'. A "don't care" in the search pattern is represented by SL = SL = '0'.

Figure 3.12b shows the CMOS-implementation of a NOR-type TCAM. The NOR-type CAM cell from Figure 3.9b is modified by connecting transistor M_4 to the $\bar{Q}$ node of the second flip-flop. Since both flip-flops can store independent information, the addional "don't care" can be generated. Figure 3.12c shows the CMOS-implementation of a NAND-type TCAM. The NAND-type CAM cell from Figure 3.9c is basically adapted. The circuit is extended by another flip-flop and further transistor M_{mask} in parallel to M_1. The gate of M_{mask} is connected the output of the second flip-flop. When this state is '1' the match line sections are joined, regardless of the state in the original flip-flop. In consequence, the NAND-type TCAM has two valid "don't care" combinations.

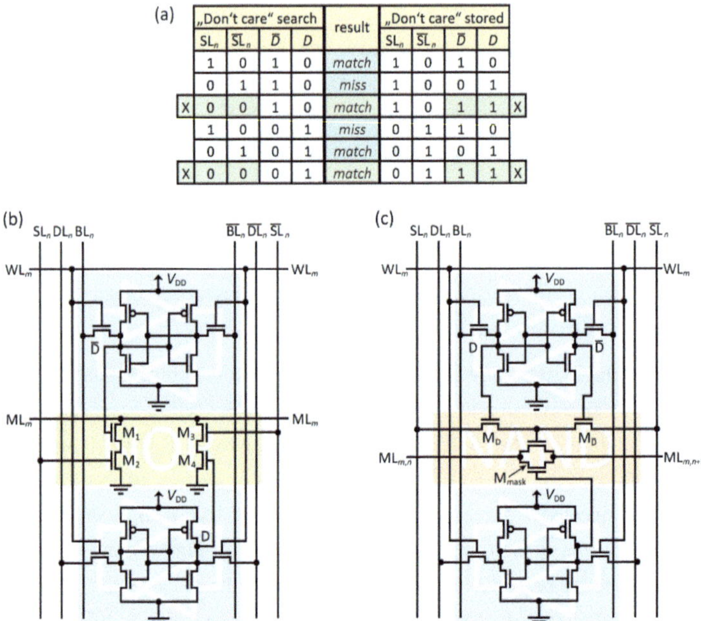

(a)	„Don't care" search				result	„Don't care" stored				
	SL_n	$\overline{SL}_n$	$\overline{D}$	D		SL_n	$\overline{SL}_n$	$\overline{D}$	D	
	1	0	1	0	match	1	0	1	0	
	0	1	1	0	miss	1	0	0	1	
X	0	0	1	0	match	1	0	1	1	X
	1	0	0	1	miss	0	1	1	0	
	0	1	0	1	match	0	1	0	1	
X	0	0	0	1	match	0	1	1	1	X

Figure 3.12: SRAM based Ternary CAM cells – (a) truth table of a CMOS-based TCAM, the "don't care" states are highlighted by the light green background (b) 16-transisor NOR-type TCAM cell and (c) 16-transistor NAND-type TCAM cell.

Summary

This Section introduces the basic idea of Content Addressable Memories to reveal the analogy between ACNs and CAMs. ACNs can be utilized to implement a CAM without the use of CMOS-technology in the memory. The basic requirements of a search iteration can be adapted to the same block diagram that is valid for CMOS-based CAMs (cf. Figure 3.8). Furthermore, by redoubling every ACN core cell, i.e. the combination of two CRS with opposite logic states (cf. Figure 3.1a), also TCAM arrays are achievable. However, to achieve the same information capacity, that requires twice the area of a standard ACN.

3.2.3 Concepts with alternative memory techniques

Concepts to optimize the current technologies were introduced by different groups. In the following some mentionable principles are shortly introduced. Ahead of this, one can summarize that all approaches do not eliminate the need of CMOS implementation in the memory array. An alternative concept based on a fully passive crossbar array will be presented in the next chapter.

DRAM

In the concept proposed by Noda [17] the TCAM core cell consisting of six transistors is extended by two preferably large capacitors. Those are mounted to the connection of the bit line to the match line transistor gate (cf. Figure 3.12). A charge stored in this position can substitute the bistable latching circuit. Like conventional DRAM, the memory is volatile and requires refreshes. However, it is considerable that this approach enables cost-efficient, large-scale, and high-performance TCAM chips for network applications.

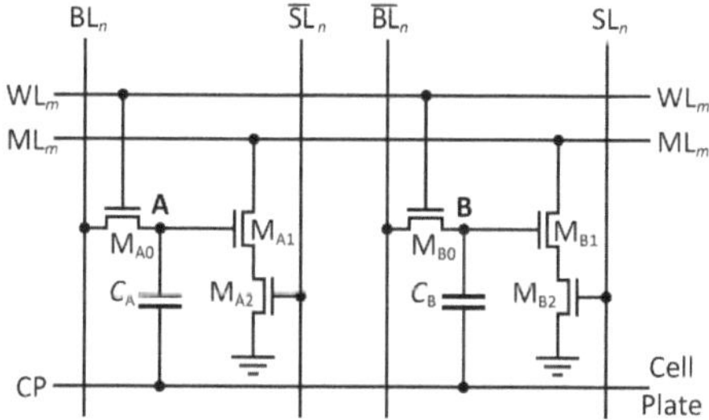

Figure 3.13: DRAM based TCAM – With the junctions A and B the TCAM states can be set to L/H and H/L for the conventional Boolean states and to L/L for the "don't care" state. Adapted from [17].

FLASH

Non-volatile implementations of TCAMs were already introduced in the 1990s by Miwa et al. and Hanyu et al. [18, 101]. Their concept implies a core cell based on a pair of single floating gate transistors sharing one match line. The idea is to exploit the programmable threshold voltage of each transistor to activate at least one transistor to discharge the corresponding match line for a non-match. Similar to conventional TCAMs, all states can be covered by this approach with this two-bit core cell. For instance, the combination of two high thresholds can represent the "don't care". The major drawback of this area saving implementation is presumably the long writing time and low endurance that makes this technique insufficient for integration in network devices.

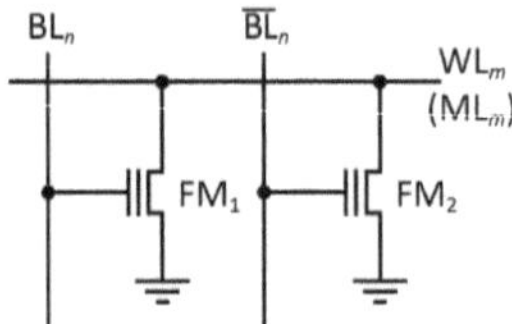

Figure 3.14: Non-volatile CAM based on two single floating gate MOSFETS. Adapted from [101].

RRAM

A large variety of resistive switch mechanisms are available to implement memories (cf. Figure 2.2). Besides ReRAMs, Magnetic Tunnel Junctions (MTJ) based on the Spin-Transfer Torque effect (STT) and Phase Change Memories (PCM) are one of the most promising candidates for future memories.

Matsunga et al. proposed numerous techniques to implement TCAM cells in a 2R2T assembly [19-20, 22-25]. As all RRAM methods can be essentially reduced to an element that has at least two different resistive states, the principle core cell is always comparable for different RRAM implementations. Like illustrated in Figure 3.15a, the series-parallel connection of two n-type MOS transistors and two RRAM devices can build the core cell. Here, the current from match line to word line can be modulated by the resistive. The 2R1T implementation in Figure 3.15b exploits that voltage dividing property of the resistive switch assembly and thereby adjusts the threshold voltage of the n-type MOSFET.

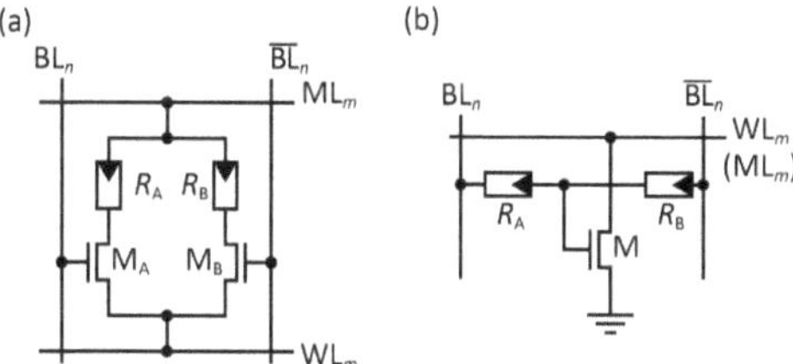

Figure 3.15: RRAM integration concepts for TCAM – (a) 2R2T [20] and (b) 2R1T [19] core cells. Adapted from [25].

Eshraghian et al. proposed a summary of assemblies that are closer based on the conventional SRAM based integration of TCAMs [26]. Like Figure 3.16 reveals, the circuit has a high analogy to the SRAM based NOR-type circuit in Figure 3.9b. Only the bistable latch circuit is replaced by two memristive elements and an additional voltage line (VL).

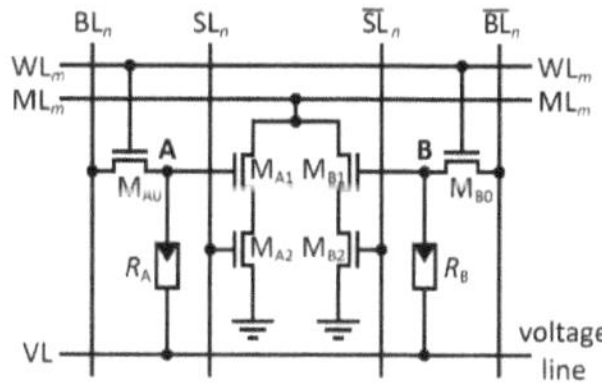

Figure 3.16: RRAM integration for a 5-T NOR-type CAM. Adapted from [26].

3.3 Simulative Classification

To study ACN array properties in detail, dynamic memristive simulation including parasitic capacitances and line resistances are conducted. The used model for memristive ECM elements (cf. Chapter 2.2) is implemented in VerilogA, and array simulations are conducted using Cadence Spectre.

3.3.1 Used Models

A physics-based memristive ECM model which offers a high degree of accuracy was used for the here conducted simulation. The physical background of the model is described in [102], and in [103] the SPICE implementation for simulation of CRS devices is discussed. Instead of SiO_2, TaO_x is considered as ion-conducting material in this work, offering a relative permittivity of 26. All other parameters are unaltered and are chosen according to [102]. ECM memory elements consist of an inert electrode (e.g. Platinum), an active electrode (Silver or Copper) and a solid electrolyte material in between. An Ag or Cu filament is formed during the SET process (positive voltage), which is dissolved during the RESET process (negative voltage). The gap width between the filament tip and the active electrode corresponds to the state variable x in the memristive modeling approach. This state variable is controlled by the ionic cell current. In the LRS the gap width x is very small, thus a large electronic tunneling current flows. In the HRS the filament is dissolved, hence only the low ionic current determines the overall current. For this study the cell capacitance of cell A and B were added, respectively, to the ECM model (cf. Figure 3.17a). A feature size of $F = 40$ nm is assumed for all simulations. Four different configurations of ReRAM-cell thicknesses for element A and B are considered.

$$
\begin{aligned}
\frac{D_A}{D_B} &= \frac{21\,\mathrm{nm}}{14\,\mathrm{nm}} = 1.5 \qquad \left(\frac{C_A}{C_B} = \frac{2}{3}\right) \\[4pt]
\frac{D_A}{D_B} &= \frac{25\,\mathrm{nm}}{12\,\mathrm{nm}} = 2 \qquad \left(\frac{C_A}{C_B} = \frac{1}{2}\right) \\[4pt]
\frac{D_A}{D_B} &= \frac{28\,\mathrm{nm}}{7\,\mathrm{nm}} = 4 \qquad \left(\frac{C_A}{C_B} = \frac{1}{4}\right) \\[4pt]
\frac{D_A}{D_B} &= \frac{30\,\mathrm{nm}}{5\,\mathrm{nm}} = 6 \qquad \left(\frac{C_A}{C_B} = \frac{1}{6}\right)
\end{aligned}
\tag{3.25}
$$

To obtain realistic array simulations, coupling capacitances ($C_{seg} = 2.76$ aF) and line resistances ($R_{seg} = 0.86\ \Omega$) were added to each segment to the simulation model of the array, see Figure 3.17. Each $4F^2$-segement consists of a CRS device connected to a 80 nm long word line section and a 80 nm long bit line section. The segment resistance and capacitance were calculated for Cu wires of height $F = 40$ nm and SiO_2 as interline material ($\varepsilon_r = 3.9$).

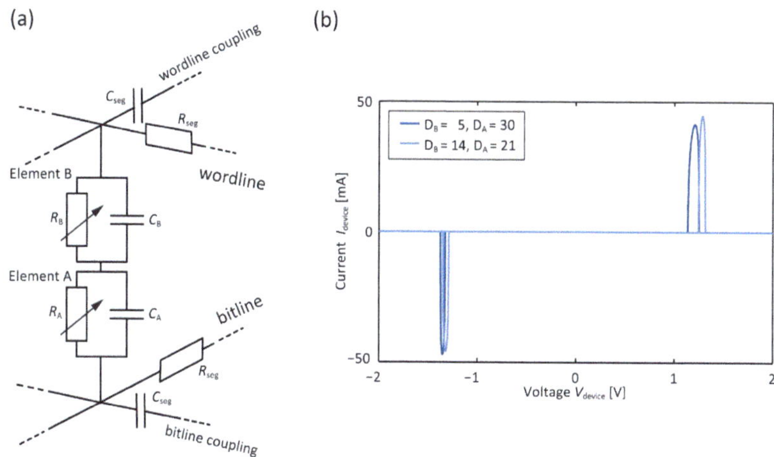

Figure 3.17: ACN simulation model – (a) one segment of the model containing the equivalent circuit diagram for a CRS device, coupling capacitances and line resistances for word and bit lines. Each ECM core element, A and B, is considered as a resistively switching element (e.g. R_A = LRS and R_B = HRS) offering an intrinsic cell capacitance C_A and C_B, respectively. (b) CRS devices I-V characteristic for the ratios C_A/C_B = 1/6 and C_A/C_B = 1/1.5.

In Figure 3.17b the simulated I-V characteristic of a single CRS device is depicted. Due to the much lower RESET voltages compared to the SET voltages in ECM devices ($|V_{RESET}| < |0.5 \cdot V_{SET}|$), the ECM-type CRS offers only spike shaped LRS/LRS states ('ON-window') [72, 103]. It is interesting to note that the positive and the negative ON-windows are slightly different in shape in Figure 3.17b. In contrast, the I-V characteristic is symmetric for completely identical ReRAM elements, A and B, cf. [103]. This difference is due to the cell asymmetry in terms of thickness. Note that the overall device characteristic does not change by introducing a cell asymmetry in terms of thickness, as can be seen for C_A/C_B = 1.5 and C_A/C_B = 6 in Figure 3.17b.

3.3.2 Array Write and Read Simulations

To enable array simulations, a bit line (bl) and a word line (wl) voltage control, a ramp signal and a comparator are added to the basic crossbar array. Figure 3.18 gives a functional overview of write and readout of a 2 × 4 array showing the corresponding transient voltages and currents of all cells. The first two lines represent the word lines and the first four columns the bit lines, respectively. Word line potentials are colored in blue and bit line potentials in red. The first tile illustrates that in the first-time step wl_0 is at +2.1 V and bl_3 is at -2.1 V. Hence, -4.2 V are applied across the cell in total which results in a switching event to state '1' (A: HRS, B: LRS). During the first step all cells are initialized with a logical '1'. In the next two steps the words '00 11' and '10 01' are stored by writing '0' (A: LRS, B: HRS) in the desired cells. At approximately $t = 0.3$ μs the write procedure is completed, and after a short down time, during which the array is discharged, the readout begins at $t = 0.4$ μs. The search pattern '10' (and the negated '01') is applied on the bit lines with a pulse height of 500 mV. The resulting word line voltages are depicted in more detail in the last tile on the bottom right along with the adjusted voltage ramp. The voltage difference causes that the comparator on wl_1 switches before the comparator on wl_0 (upper right tiles) indicating that the search pattern matches with the data on wl_1. The other four tiles in the bottom line represent the current on each bit line. Evidently, the write operation currents are significantly higher than the currents during the search operation. However, the write operation is a relatively rare event compared to search operations.

	Informational word						Negated word					
	MSB					LSB	MSB					LSB
	bl_{2n}	bl_{2n-1}	bl_{2n-2}		bl_{n+2}	bl_{n+1}	bl_n	bl_{n-1}	bl_{n-2}		bl_1	bl_0
wl_0	0	0	0	⋯	0	0	1	1	1	⋯	1	1
wl_1	1	0	0	⋯	0	0	0	1	1	⋯	1	1
wl_2	1	1	0	⋯	0	0	0	0	1	⋯	1	1
	⋮	⋮	⋮	⋱	⋮	⋮	⋮	⋮	⋮	⋱	⋮	⋮
wl_{n-1}	1	1	1	⋯	0	0	0	0	0	⋯	1	1
wl_n	1	1	1	⋯	1	0	0	0	0	⋯	0	1

Table 3.2: Structure of truth table for simulations – the incremental structure effects that each stored word has the HD 1 to its neighboring word, e.g. line wl_3 has the HD = 1 to wl_2 and wl_4, HD = 2 to wl_1 and wl_5, etc.

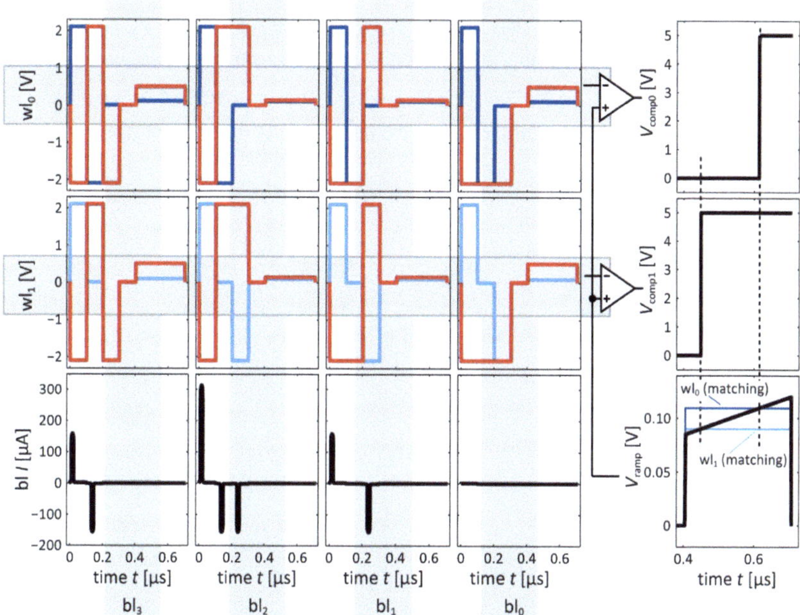

Figure 3.18: Exemplary 2×4 array simulation – here, the capacitance ratio of $C_A/C_B = \frac{1}{2}$ is used. Left: For each of the cross point junctions of the 2×4 array the word line potential (wl$_0$: blue, wl$_1$: light blue) and the bit line potential (red) is depicted. Moreover, the corresponding bit line currents (black) are shown for each bit line. Bottom right: The comparator inputs during the searching operation are depicted: the ramp (black), the wl$_0$ signal (blue) and the wl$_1$ (light blue). If the ramp signal is larger than the match line voltage, the comparator output switches to 5 V (see graphs on the right for V_{comp0} and V_{comp1}).

3.3.3 Implementation of a $N \times 2N$ array

For all simulations a $N \times 2N$ array offering a regular structure according to Table 3.2 is considered. The information part of the first word line (wl$_0$) only contains '0' for all considered arrays. In the next word lines the Hamming distance is increased successively by alternating the first '0' to '1'.

To illustrate the pattern matching functionality, we consider an 8×16 array constructed according to Table 3.2. The first search pattern is '0000 0000'. In Figure 3.19a, the comparator output for each word line is depicted. The input signal of a '1' is represented by 'pulse'; in this example, it is only applied to the negated word.

As expected, wl_0 is the first line which pulls up to 5 V, indicating a full match. Due to the regularity of the considered matrix, wl_1 to wl_7 respond successively. As a second example the search pattern '1110 0000' is examined (see Figure 3.19b). In this case wl_3 is the full match and the lines wl_2 and wl_4 offer a Hamming distance of HD = 1, whereas wl_1 and wl_5 offer HD = 2. The least similar pattern for this search is stored in wl_7 (HD = 4).

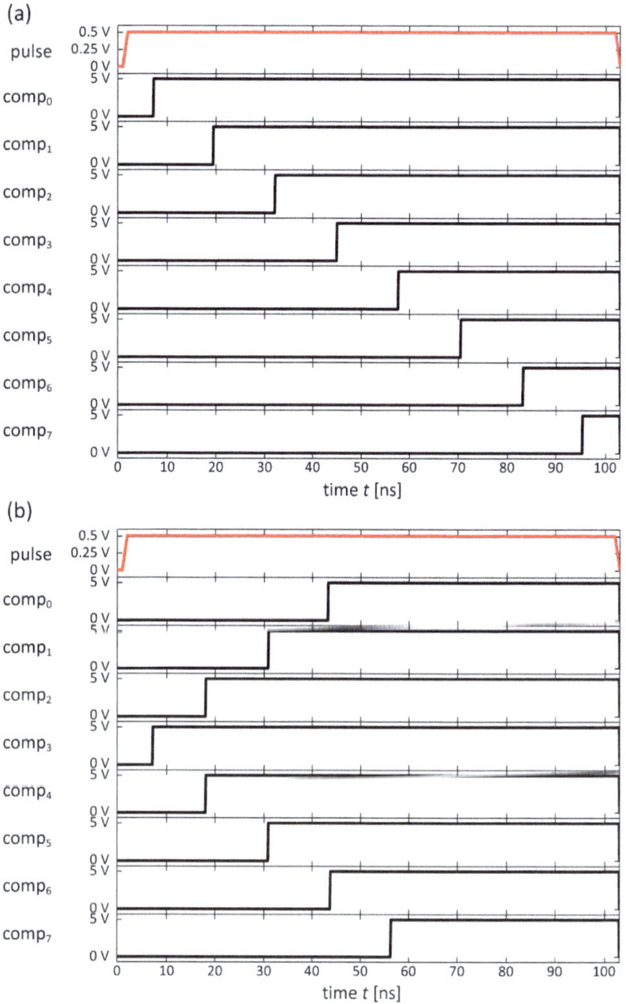

Figure 3.19: ACN signal simulations for a 8×16 array – (a) simulations for the search pattern '0000 0000' which leads to the first output on wl_0 and (b) for search pattern '1110 0000' which results in an initial detection on wl_3. In this simulation $C_A/C_B = 1/2$ is considered.

3.3.4 Minimum Voltage Margin

To study the minimum voltage margin ($\Delta V = |V_{HD0} - V_{HD1}|$), simulations on 2×4 and 4×8 arrays for all possible search patterns were conducted (cf. Figure 3.20a,b). Depending of the stored information within the array, the achievable voltage distance ΔV between a pattern with HD = 0 and HD = 1 can vary. For the 2×4 array '10' is the worst-case search pattern leading to a match, whereas for the 4×8 array there are two patterns ('1000' and '1100') leading to matches showing the smallest ΔV.

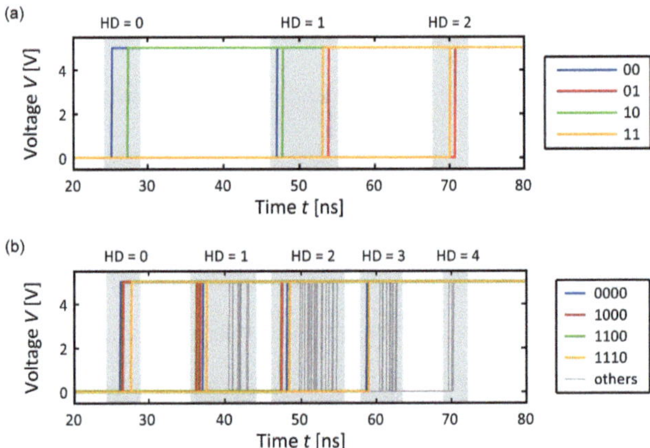

Figure 3.20: (a) Summary of all search pattern in a 2×4 array and (b) 4×8 array (search pattern with HD $\neq 0$ are gray colored) in the same time domain. For these simulations $C_A/C_B = 1/6$ is considered. Regions for each HD are marked by gray background color.

From Figure 3.20b one can see that the voltage margin begins to shrink for Hamming distances that often occur. Nevertheless, the search time is a qualitative indicator of similarity. However, if just a full match is requested, $\Delta V = |V_{HD0} - V_{HD1}|$ is the only relevant figure of merit. In Figure 3.21 ΔV is depicted for several array sizes $N \times 2N$ up to 32×64 and different capacitance ratios $C_A/C_B = 1/1.5$, $1/2$, $1/4$, and $1/6$, respectively. We consider search patterns containing $N/2$ '1' followed by $N/2$ '0', i.e. '11110000' for the 8×16 array. For $C_A/C_B = 1/6$ the minimum voltage margin is $3.4\,\text{mV}$, whereas for $C_A/C_B = 1/1.5$ the lowest $|V_{HD0} - V_{HD1}|$ is $0.7\,\text{mV}$ in a 32×64 array. The lowest voltage difference is determined by the performance of the sensing circuitry. Values in the range of several hundreds of microvolts can still be covered by discrete circuit elements (e.g.

the LMP7300 by Texas Instruments or the MAX995 by Maxim Integrated). Fully integrated circuits should be capable of even higher sensitives.

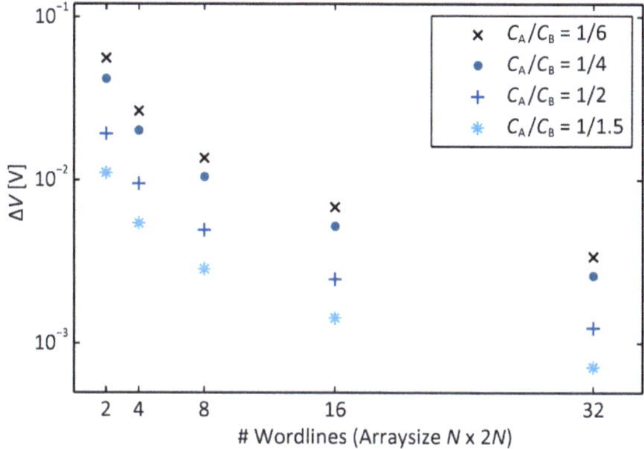

Figure 3.21: Matrix dimension N of a $N \times 2N$ array versus minimum voltage margin ΔV for different ratios of C_A to C_B.

3.3.5 Energy considerations

To study the energy consumption, the worst-case search pattern is evaluated first. All considerations within this section were made for an ACN where '0' are represented by large capacitances and '1' by small ones. Also the readout scheme was determined by case A of Table 3.1, thus, a high applied voltage represents a '1' and no voltage represents a '0'. Since CRS devices being in state '0' offer a larger overall capacitance compared to state '1' in general, the total number of accessed CRS devices in state '0' is the figure of merit for the overall energy consumption. In Table 3.3 the application of the search pattern '00001111' for an 8×16 crossbar array is exemplified. All accessed CRS devices (i.e. the devices where a voltage is applied) ΔV in state '0' are marked by gray background color and the sum of zeros in each column is stated. The total sum is 48, which is the worst case for this type of array. Thus, a search word containing $N/2$ zeros followed by $N/2$ ones is the worst-case search pattern in terms of energy. This type of search pattern was applied for the energy calculation of arrays of different sizes and $C_A/C_B = 1/1.5$, $1/2$, $1/4$, $1/6$ (cf. Figure 3.22).

For a 32×64 array and a $C_A/C_B = 1/1.5$, a minimum energy of 4.69 aJ/bit/search ($= W_{min,32\times64} / (\#wl \cdot \#bl/2)$) is feasible. For a 32×64 array and a $C_A/C_B = 1/6$, the search energy is 9.78 aJ/bit/search. The graph shows that the smallest capacitance ratio offers the lowest energy consumption, here. This result highlights an intrinsic tradeoff between achievable voltage margin and minimum energy. However, in most cases the voltage margin will be the most relevant optimization goal for ACN circuits, as can be seen from the minimum voltage margin study (cf. Figure 3.21). Thus, larger C_B values compared to C_A are nevertheless desirable for implementation of larger array sizes.

	Informational word								Negated word							
	A	B	C	D	E	F	G	H	$\bar{A}$	$\bar{B}$	$\bar{C}$	$\bar{D}$	$\bar{E}$	$\bar{F}$	$\bar{G}$	$\bar{H}$
A	0	0	0	0	0	0	0	0	1	1	1	1	1	1	1	1
B	1	0	0	0	0	0	0	0	0	1	1	1	1	1	1	1
C	1	1	0	0	0	0	0	0	0	0	1	1	1	1	1	1
D	1	1	1	0	0	0	0	0	0	0	0	1	1	1	1	1
E	1	1	1	1	0	0	0	0	0	0	0	0	1	1	1	1
F	1	1	1	1	1	0	0	0	0	0	0	0	0	1	1	1
G	1	1	1	1	1	1	0	0	0	0	0	0	0	0	1	1
H	1	1	1	1	1	1	1	0	0	0	0	0	0	0	0	1
	0	0	0	0	1	1	1	1	1	1	1	1	0	0	0	0

# of zeros $5 + 6 + 7 + 8 + 7 + 6 + 5 + 4 \quad = \quad 48$

Table 3.3: Maximum energy search pattern – Regarding power consumption, the worst case search pattern is highlighted by red background color. All accessed CRS devices in state '0' are marked by gray background color. The sum of zeros in each accessed column is given. The total sum is 48 in this case.

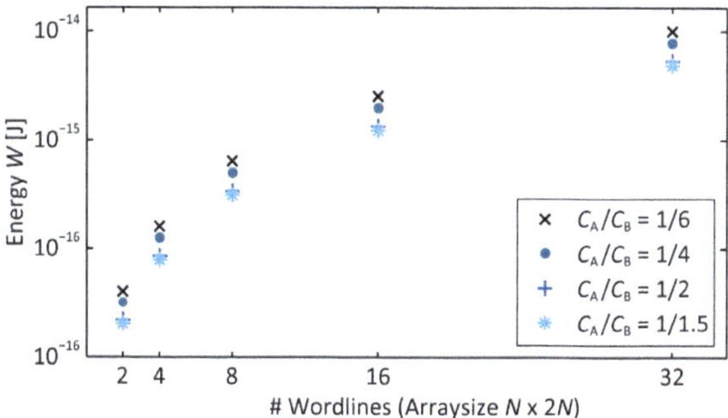

Figure 3.22: Matrix dimension N of a $N \times 2N$ array versus energy consumption W for different ratios of C_A to C_B.

3.3.6 Large Array Calculations

The simulations of large arrays are carried out using HSPICE. A nonlinear approach to model CRS devices was used [104, 105]. In this test the stored patterns are prepared to provide a complete range of Hamming distances, from $HD = 0$ to $HD = M$. Therefore, the array size for storing templates and their negates is $(N = M + 1) \times (2M)$. The nomenclature from Chapter 3.1 is still valid. According to the cell description T_{ij} it is assumed that $i \in \{1, 2, \ldots, N\}$. In consequence, the stored and search patterns are completely matched for $i = 1$ and there is a complete mis-match where $i = N$. This is shown in Figure 3.23, where $M = 40$ and $N = 41$. $M = 40$ is equivalent to the minimum commercially available search key-width of Renesas Electronics' ternary CAM that is applicable in enterprise switches and routers, 3G/4G mobile access platforms, IPv4 and IPv6 packet forwarding and Internet protocol security (IPSec) [106].

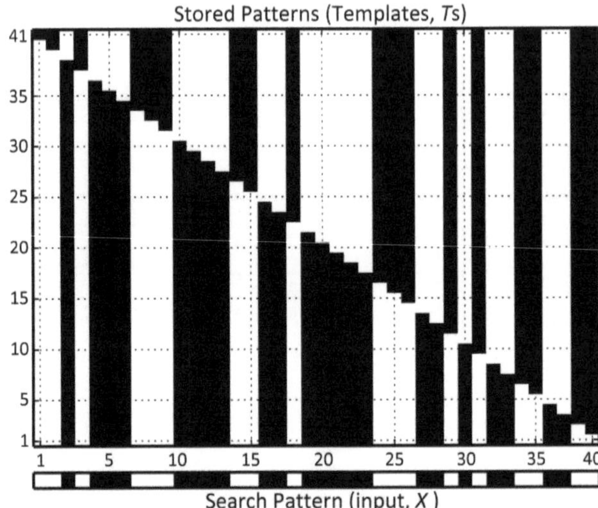

Figure 3.23: Stored pattern in the 41 × 40 test-bench nano-crossbar array. The complementary array is not shown here. Search pattern is also demonstrated. The search pattern has HD = 0 with stored pattern 1 and maximum distance (HD = 40) with stored pattern 41. Black and white pixels represent '0' and '1' bits, respectively. For the search pattern, X, a white pixel shows that a pulse exists on that input, x_j, and a black pixel shows that the pulse does not exist on x_j. Adapted from [91].

The crossbar array is fitted with reset (initialization) circuitry, as already described in Figure 3.1a by S_1, S_2 and S_3. These switches are closed when the evaluation phase is disabled. During the evaluation phase EVAL, N match-lines (MLs) are read-out and, as demonstrated in Figure 3.24, a clear separation between voltage ranges is observed. The MLs' voltage range is between 0.51 V to 1.54 V. Here the first output has HD = 0 and 41th output contains a pattern with maximum HD = 40 to the input bit-vector. The output that is represented in Figure 3.24 is consistent with Eq. (3.13) for $C_A = 0.3$ pF, $C_B = 0.9$ pF ($c_r = 3$), $\Delta V_x = 3$ V and $C_{\text{ML},i} = 22$ pF. These data are extracted from NDRO experiments in Section 2.4.2 [79].

For the characterization of the ACN, the intra Hamming Distance (intra-HD) is introduced, $|V_{\mathrm{ML},i} - V_{\mathrm{ML},k}|$, where $k \in \{i+1, ..., N\}$, which represents the Hamming distance between two distinct stored patterns. Each intra-HD corresponds to a certain voltage difference. For N match-lines, the total number of paired line comparisons can be calculated as

$$\text{Total number of pairs} = \binom{N}{2} = \frac{N!}{(N-2)! \, 2!} \tag{3.26}$$

The number of output pairs represents the total count of ways to choose $k = 2$ lines from a set of $N = 41$ lines. For $N = 41$, the number of output pairs is 820. In Figure 3.25, the y-axis shows the output pair index and the absolute voltage difference is shown on the x-axis. For identification the intra-HD with their corresponding number of output pairs is listed in Table 3.4. Note that the table's value for $|V_{\mathrm{ML},i} - V_{\mathrm{ML},k}|$ highlights the ability to distinguish between two V_{ML} outputs. The variation in the reference voltage V_{th} and the comparator off-set voltage mismatch contribute to detectability of the nearest-match, however, here we assume that readout circuitry is ideal and V_{th} is stable. As discussed earlier, this circuitry is a test setup to analysis the ACN's functionality. The inset in Figure 3.25 shows a variation in readout voltages for the same HD match. This is due to the impact of parasitic elements, such as nano-wire segment resistors $R_{\mathrm{seg}} = 1.25\ \Omega$, for each memory cell.

Intra-HD	Number of pairs	$|V_{\mathrm{ML},i} - V_{\mathrm{ML},k}|$
1	40	24.5 mV
2	39	49.0 mV
3	38	73.5 mV
4	37	98.0 mV
5	36	122.0 mV
6	35	147.0 mV
7	34	171.5 mV
8	33	196 mV
...	...	...
39	2	956.5 mV
40	1	981.0 mV

Table 3.4: Intra-HDs and their corresponding number of output pairs

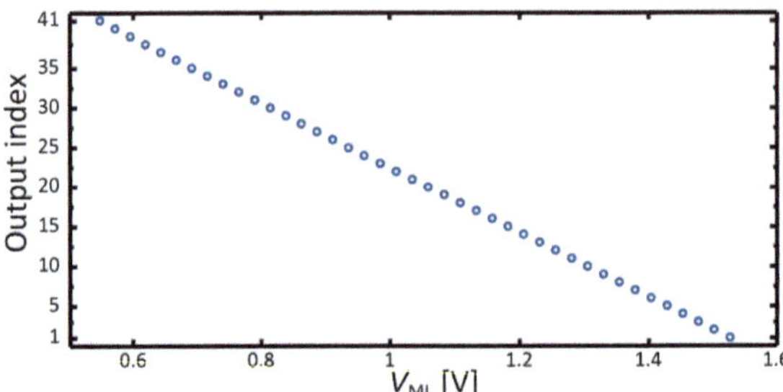

Figure 3.24: Nanocrossbar array's output voltages of an ideal network. The stored patterns match to the example in Figure 3.23. Adapted from [91].

The above-mentioned assumptions allow to perform a Monte Carlo analysis on a 41×40 ACN. In this simulation the independent variables of all cells were randomly picked and fed into the deterministic HSPICE simulation. Variables for the Monte Carlo simulation are highlighted in Table 3.5. The distribution that is assigned to all random variables' is Gaussian with relative 3σ deviation which is also mentioned in Table 3.5. These values are applicable for the investigated large-scale µm cells. In this simulation local and global variation between similar components within one array (local) and across similar arrays (global) are considered. A Monte Carlo simulation with (more than) 7000 runs is performed and the result is summarized in Figure 3.26.

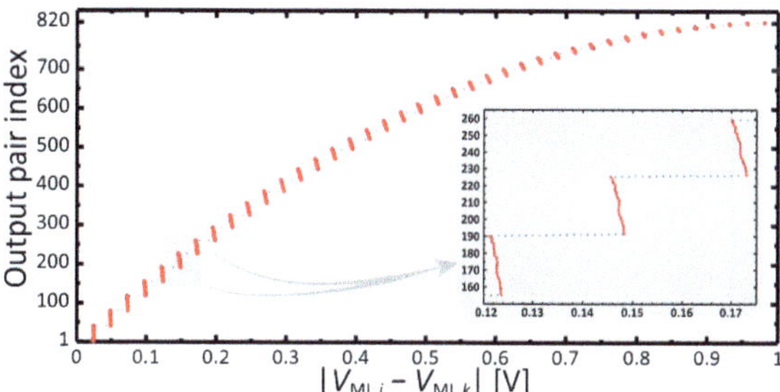

$$|V_{ML,i} - V_{ML,k}|\ [V]$$

Figure 3.25: All possible intra-HDs in the 41 × 40 array – the y-axis shows the index of all possible intra-HD combinations. Values on the x-axis show the absolute voltage difference between the output voltage of the individual pattern pair. Patterns are sorted, so that intra-HD = 1 is represented by the first step (cf. Table 3.4, line 1), intra-HD = 1 by the second step (cf. Table 3.4, line 2), and so on. The inset shows the voltage variations for one intra-HD step. It demonstrates the impact of parasitic elements, such as nano-wire segment resistors for bit- and word-lines, which are considered in simulations. Adapted from [91].

Variable	Mean (μ)	Relative 3σ
Supply and input voltages	3 V	10 %
Series resistors on each ML	100 Ω	10 %
Device thickness	20 nm	10 %
Top electrode width	5 μm	10 %
Middle electrode width	10 μm	10 %
Bottom electrode width	15 μm	10 %
Load capacitor (C_{ML})	22 pF	10 %
ON resistance(R_{ON})	1 kΩ	20 %
OFF resistance(R_{OFF})	1 MΩ	20 %

Table 3.5: Independent random variables for Monte Carlo simulation

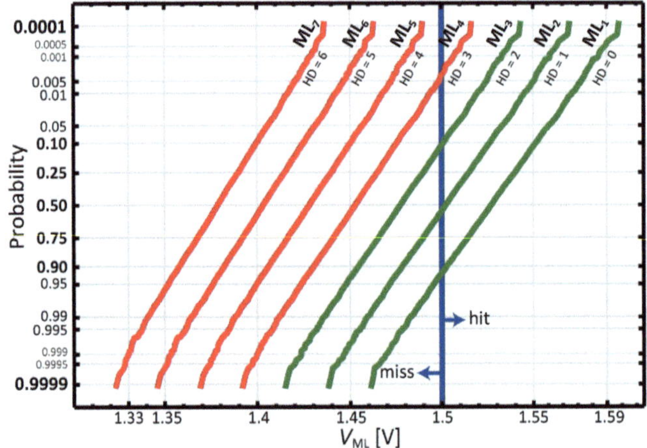

Figure 3.26: Probability-voltage distribution for seven outputs with HD = 0...6 that correspond to ML_1 to ML_7. The solid blue line indicates V_{th} in this test-bench circuit. Outputs with a voltage amplitude above the threshold voltage are treated as 'hit' and below that are treated as 'miss'. Adapted from [91].

To have a clear understanding of the detection capability of the simple readout circuitry, Figure 3.26 illustrates the probability function p of the 7 top match-lines with minimum HDs versus V_{ML}. The detection probability can be interpreted as p. In order to detect HD = 0 (ML_1) with $p = 95\,\%$ chance, V_{th} has to be around 1.5 V. Under these circumstances a successful detection over a range of outputs is achieved. If HD = 0, the output is 1 with 95 % probability. For HD = 1, the output is 1 with 50 % probability, while for HD = 2 the probability for an output 1 is 10 %. For HD = 3, only a probability of 0.5 % for observing an output 1 is given. The probability that output is 1 for HD > 3 tend to negligible small values. Thus, HD < 4 are detectable with high probability, i.e. the found match has a HD < 4 to the search pattern with almost 100 % probability.

3.4   Experimental

Experiments on real samples are conducted to verify the computational simulations. Therefore, an experimental test setup composed of SMD circuity boards (cf. Figure 3.27a) with external function generators was built and microstructure arrays of pre-programmed capacitive devices (cf. Figure 3.27b) for validation were fabricated [107,

108]. In the following, the test setup, the fabrication of the samples and the subsequent measurements are described.

3.4.1 Test Setup

The experimental test setup was developed to reproduce the basic features of ACNs that were implemented in the simulated circuity in the previous section. The setup contains a circuit board for impedance conversion; a voltage-to-time-conversion and a power supply for an 8×16 array. A Winner-Take-All (WTA) [109] function with sample and hold element and a counter for the Hamming Distance are optionally available. The excitation pulse and the voltage ramp for voltage-to-time-conversion can be applied via external function generators.

To prevent influences of noise the impedance converters (cf. Figure 3.28) are mounted in proximity to the device under test (DUT). The low-impedance output allows further processing, in this case the conversion from its voltage level to time. This is achieved by comparing the signal with a defined voltage ramp using a suitable operational amplifier. Depending on the respective HD of each WL the signal has a distinct voltage level. Since the spectrum of levels is relatively small (cf. Figure 3.21) it can be hard to distinguish two specific levels. The voltage-to-time-conversion allows a much more precise HD detection.

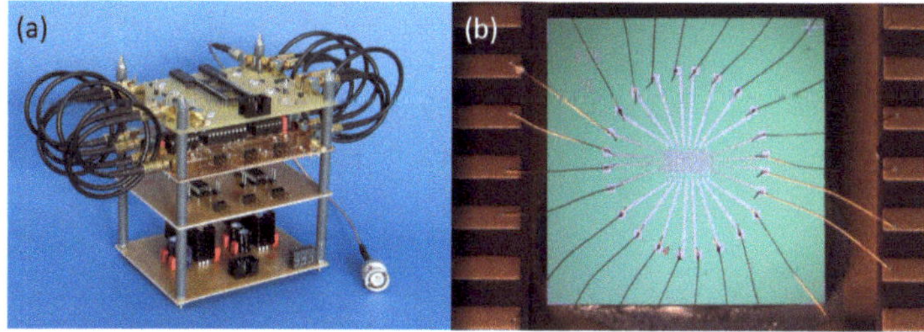

Figure 3.27: Test setup – (a) circuits with SMB and BNC coaxial connectors. The lowest level is the power stabilization, the intermediate levels are for evaluation support. The core of the read-out scheme is based on the top level, where also the DUT is mounted. (b) microscope photo of a 8×16 μ-structure ACN array bonded to a 28-pin chip carrier.

The feasibility study was performed using technologically preprogrammed μ-meter test samples (cf. Figure 3.27b). The 8×16 arrays were fabricated on wet-oxidized 1″ silicon substrate (450 nm SiO_2) and a thermally oxidized 5 nm TiO_2 adhesion layer. The platinum word lines (30 nm thickness) and bit lines (50 nm thickness) were fabricated using conventional UV-lithography for lift-off structuring and DC magnetron sputtering for deposition. The planar TiO_2 isolation layer of 30 nm thickness was manufactured by high pressure reactive DC sputtering in a 74 % argon and 26 % oxygen atmosphere. The dielectric permittivity ε_r of 23 in the deposited TiO_2 is reasonable high to promote the capacitive readout [86]. Consistent to the different memory states the manufactured cell areas are either $24 \times 10 \ \mu m^2$ or $20 \times 20 \ \mu m^2$. According to the 20 nm TiO_2 layer the effective cell capacitance is either 2.45 pF for state '1' or 4.1 pF for state '0'.

The arrays are separated in $5 \times 5 \ mm^2$ dies which are mounted in a 28 pin DIL chip carrier. The contact pads for the 24 word- and bit lines are bonded to the carrier pin outs using gold wired wedge-wedge bonding.

The pulses are generated using an Agilent 81110A Pulse Generator, whose signals were split to the search pattern by 16 mechanical on-board switches. The voltage ramp for the voltage-to-time-conversion is supplied by a Hewlett Packard 3314A Function Generator, which is triggered by the pulse generator. The evaluation of the output signals is performed by a 4 channel Tektronix oscilloscope model DPO73304D with two P6139B high impedance voltage probe heads.

The basic schematic of the developed circuit is depicted in Figure 3.28. Search patterns are generated manually by simple mechanical on-off switches S_0–S_{15}. Switches S_0–S_7 build the informational search pattern while S_8–S_{15} generate its inverse. That means that the user is responsible for the correct negation of the search pattern, which is acceptable for this experimental purpose. The resulting weighted word line signal is then fed into the impedance converter (cf. Figure 3.28b), which is supported by an OPA656N operational amplifier. The converted signal is then applied to the comparator, model MAX944CSD+.

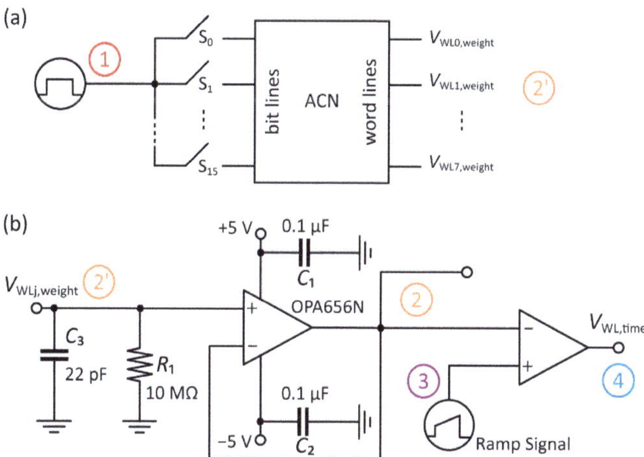

Figure 3.28: ACN evaluation circuit schematic – (a) Point ① indicates the excitation pulse which forms the search pattern by the switches S_0–S_{15}. Point ②′ indicates the weighted results on each word line. (b) Each word line is connected to an impedance matching where the weighted voltage $V_{WLj,weight}$ is converted (point ②) and is then compared by an operational amplifier with a voltage ramp (point ③). The resulting voltage $V_{WL,time}$ is indicated with point ④. The marking points ①–④ can be retrieved in Figure 3.29.

3.4.2 Test Results

For the highest practical value the focus was only directed on one specific set of patterns that corresponds to the truth table shown in Table 3.6. The table, which can be derived from Table 3.2, is sufficient for investigations on ACNs since all words cover the intrinsic HDs from 1 to 7 but are very easy to separate due to their increasing structure, as introduced in Section 3.3.3. That simplifies understanding of results for one specific search pattern and makes its analysis more manageable.

	Informational word								Negated word							
	A	B	C	D	E	F	G	H	$\bar{A}$	$\bar{B}$	$\bar{C}$	$\bar{D}$	$\bar{E}$	$\bar{F}$	$\bar{G}$	$\bar{H}$
A	0	0	0	0	0	0	0	0	1	1	1	1	1	1	1	1
B	1	0	0	0	0	0	0	0	0	1	1	1	1	1	1	1
C	1	1	0	0	0	0	0	0	0	0	1	1	1	1	1	1
D	1	1	1	0	0	0	0	0	0	0	0	1	1	1	1	1
E	1	1	1	1	0	0	0	0	0	0	0	0	1	1	1	1
F	1	1	1	1	1	0	0	0	0	0	0	0	0	1	1	1
G	1	1	1	1	1	1	0	0	0	0	0	0	0	0	1	1
H	1	1	1	1	1	1	1	0	0	0	0	0	0	0	0	1

Table 3.6: Truth table for the experimental 8 × 16 test array – the incremental structure effects that each stored word has the HD 1 to their neighboring word, e.g. line D has the HD = 1 to line C and E, HD = 2 to line B and F, etc.

The characteristic signals of the circuitry are depicted in Figure 3.29. The first line represents the pulse input that resembles a logical '1' in the applied search pattern. Here, a pulse height of 2 V was chosen as it is a sufficient voltage for investigations, but does not switch most ReRAM cells during short pulse lengths due to the non-linearity of the switching kinetics [64]. The summarized signal of one representative word line is displayed in orange in the second line of Figure 3.29. Depending on the HD it varies in the range of 0.2–0.4 V when the search pattern is applied. The purple signal in the same line illustrates the applied voltage ramp. The last line shows the output voltage of the comparator. As one can see the state is changed from 0 V to 5 V when the voltage ramp exceeds the word line signal. The red bar indicates this moment. The noise in the signals is caused by additional cables that are necessary for the performed measurements. The signals on the board or on the chip are less affected by noise since they are much closer connected to each other. Nevertheless, electromagnetic shielding will certainly bring some improvement. Moreover, the used external ramp function leaves a margin for further improvement by better integration. However, the results are repeatable and reliable, thus enabling proper array testing.

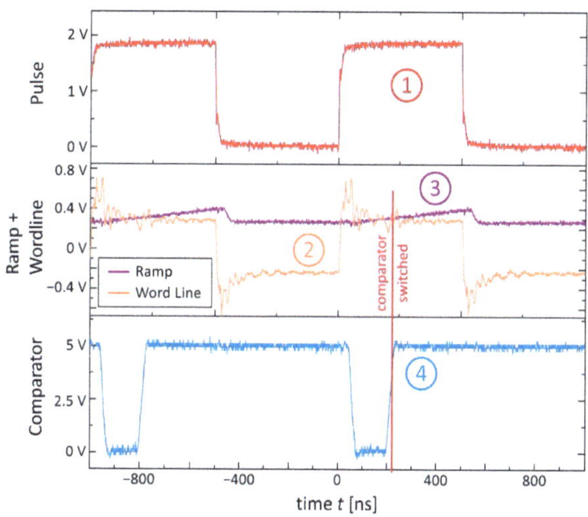

Figure 3.29: Typical signals of the test circuit – the first line (point ①) depicts the excitation pulse which forms the search pattern. In the second line the weighted voltage $V_{\text{WL,i,weight}}$ (point ②) and the voltage ramp (point ③) are shown. They form the voltage-to-time conversion in the last line (point ④). Points ①–④ correspond to the marks in Figure 3.28.

To show the capability of ACNs to find the searched pattern as well as the feature of determining the HD of the search pattern with respect to all stored patterns, the comparators on all word lines are traced at the same time while one specific search pattern is applied. As depicted in Figure 3.30 word line voltages pull up at a certain point in time, which corresponds to the HD of the stored pattern with respect to the search pattern. For the first case (cf. Figure 3.30a), where the search pattern '0000 0000' is applied, comp_0 switches after $t_{\text{wl0}} \approx 200$ ns since the match reveals the lowest HD, see Table 3.6. The word line wl_1 has a HD = 1 for this search pattern, thus comp_1 offers the second switching event at $t_{\text{wl1}} \approx 250$ ns. Finally, wl_7 which has the highest HD from the search pattern causes comp_7 to switch ON at $t_{\text{wl7}} \approx 550$ ns. All HDs can be easily determined by the switching order of the comparators of each word line, and the typical switching delay is $\Delta t_{\text{HD}} = 25$ ns. Figure 3.30b confirms the previously described behavior as the system works for arbitrary search patterns like '1110 0000', too. This information is also stored in wl_3 of the chip. In this case two word lines offer the same HD: wl_2 and wl_4 show HD = 1, wl_1 and wl_5 offer HD = 2, and for wl_0 and wl_6 a HD = 3 is obtained. Slight variations in the switching time can be observed in the measurements, see Figure 3.20. Accordingly, for the recognition of all HDs a certain time window for each HD needs to be defined.

In summary, the measurements show that the simulative predicted behavior can be observed in capacitive arrays. Pattern matching within the margin of hundreds of nanoseconds could be performed – a result which reaches the limits of the utilized test equipment, but has potential for optimization by on-chip integration. If a complete HD determination is not needed and the memory reveals no duplicated words, a WTA circuitry (cf. Appendix 2) can be used to detect only the first match. After the time step corresponding to $HD = 0$ the procedure can be terminated. Therefore, the search time can be further reduced.

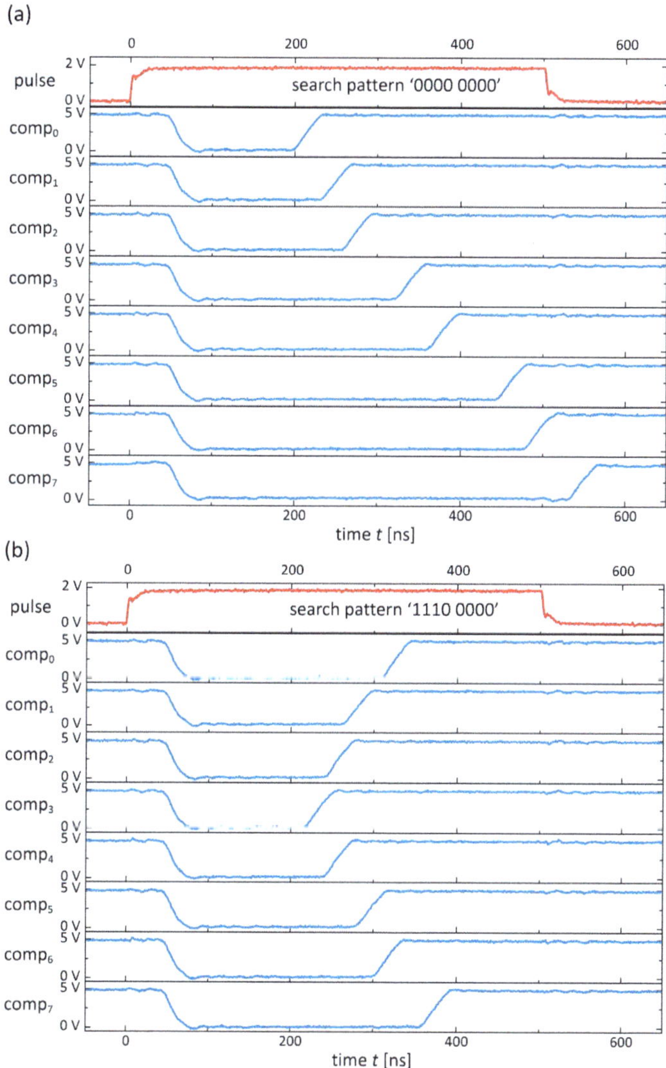

Figure 3.30: ACN measurements in the nanosecond regime – (a) ACN measurement of the search pattern '0000 0000' and (b) '1110 0000'. Corresponding to Figure 3.19 the word line wl_0 and wl_3 are detected as the best match since they are switched initially.

3.5 Discussion

The simulations show that with increasing array size the voltage margins for HD detection are decreased. The main reason is the inevitable parasitic capacitances in such an array. There are several strategies to improve the voltage margin. First, larger voltages could be applied. This is a valid approach for memristive devices offering ultra-steep switching kinetics, but it requires very short pulses to avoid unintended switching. Another option is to increase the capacitance ratio of element A and B. For example, DRAM-like trench capacitor structures could be adopted to implement very large capacitance ratios. However, for $4F^2$ crossbar arrays a factor of 6, as assumed in this work, is a realistic value. In terms of energy consumption, ECM-type CRS devices offer values below 0.01 fJ/bit/search enabling highly scalable and energy-efficient CAMs. ECM cells offer low leakage currents and exhibit a very high resistive OFF-state. That makes them favorable for ACNs compared to VCM-type CRS cells. In contrast to state-of-the-art CMOS-based CAMs, the ACN features nonvolatile information storage and a non-destructive readout mechanism. Besides the already described low power consumption this is further advantage concerning stand-by power consumption. The experimental results from micro-scale test devices confirm the ACN approach and the simulations show that also nano-scale CRS arrays can work as associative memories properly.

An ideal ACN provides first-match and multi-match detections fairly straightforward, therefore the proposed architecture is applicable in a wide range of applications including new network applications, such as intrusion detection systems, that require information about all the matching results and there is no priority among the results [110, 111]. More importantly, significantly higher integration density and the prospects of extreme energy efficient systems based on CRSs, make the ACN an attractive alternative for currently available binary and ternary CAMs. As a rule of thumb, this hardware solution is applicable, where search time is the major issue and therefore pure software solutions cannot be used.

In the following advantages of ACNs are listed:

- ACNs are feasible by a fully passive 2-CRS cell implementations for nano-crossbar arrays. Accordingly, two transistors are enough to address one stored data information in one column (including the negate). Conventional CAM structures require separated write, preload and search lines, which imposes a limitation over area, integration, scalability and power specifications. Furthermore, those transistors require a fully ON state during writing and fully OFF for data retention. In ACNs those restrictions are more relaxed.

- ACNs only require two CRS cells for one data bit. SRAM based implementations require 10 to 16 transistors per cell (cf. Section 3.2.2).

- Like every RRAM based memory, ACNs are non-volatile memories. Hence, there is no requirement for refresh cycles like in conventional associative charge-based memories.

- The cell capacitance of a non-destructive-read-out CRS cell is primarily depending on the physical cell dimensions. Thus, the actual stored resistance has just a small impact on the accuracy of the technique. As lithography is well controlled, it can be expected that the accuracy of ACNs is higher than in other resistance-based CAM implementations (cf. Section 3.2.3)

3.6 Conclusion

In this chapter ACNs were introduced and explained. It was shown that the non-destructive readout enables CRS devices for content-addressable memory functions.

It shows the feasibility of reconfigurable associative capacitive networks by means of dynamical memristive array simulations and experimental investigation of pre-programmed test arrays. For this, a reliable test setup was developed and basic pattern matching capability was proven on a nanosecond time scale. Furthermore, accurate dynamical simulations for arrays of $F = 40$ nm using ECM-type CRS device models were conducted. By considering realistic parasitic capacitances and line resistances, it can be shown that the energy consumption of a 32 bit ACN array is less than 0.01 fJ/bit/search. Monte Carlo simulations on larger array sizes up to 41×40 have demonstrated feasibility for a meaningful variation of environmental variations.

The technique of ACNs reveals a lot of opportunities for brain-inspired computation methods and has the potential to be an important step towards future neuromorphic computers.

4 Fuzzy logic

Apart from the previously described digital approach for pattern recognition, fuzzy logic computing has recently gained relevance. In the last few decades analog computing was disregarded due to lack of precision as well as the difficulty to configure analog systems. Now, as the integration of digital circuits is facing physical limits, analog systems might find new applications since they could provide small sizes and low power consumption [112].

Resistively switching devices, could be the key enabler for this new kind of analog computing since they combine non-volatility, low-voltage operation, excellent scaling properties and a wide range of multilevel states [4-5, 44, 113-114].

Another emerging trend is to use novel parallel hardware architectures to solve problems where large amounts of data shall be processed [4]. Again, resistively switching devices are considered as a highly interesting solution to enable such parallel computing paradigms [113-116].

A highly important computational task, which is required in many application domains such as data analysis and signal processing, is sorting. Today, those tasks can be most efficiently implemented in parallel digital hardware, e.g. graphical processing units (GPUs) [117, 118] or field programmable gate arrays (FPGAs) [119]. However, in terms of energy-efficiency and area consumption, a RRAM-based sorting network would be advantageous [120]. A number of concepts for sorting networks via memristive systems have been suggested recently, e.g. a cellular automata inspired approach was proposed [121] or the approach via minimum and maximum gates [122, 123]. The latter approach was implemented within this work.

In the following sorting networks with ReRAM-based comparators are described and analyzed. To this end, a device-level implementation of a minimum and maximum gate is introduced and demonstrated first. Later, a sorting network approach based on such gates is analyzed. Feasibility by real device behavior of resistively switching elements is analyzed and challenges and advantages are highlighted.

4.1 Overview

In conventional Boolean logic a statement can be true or false – only two defined states are possible. However, human thinking allows additional intermediate states. The classical logic concept can be adapted to human like thinking by blurring the results to states like 'almost true' or 'rather false'. This is inspired by situations which cannot be clearly matched to an explicit statement, i.e. result (e.g. quite expensive, rather new, very late, a little old, etc.). Allowing intermediate states expands the system to the field of fuzzy logic [124, 125].

Prominent examples for fuzzy cases are ranges of temperature, the age of people or fuel consumption of cars. For instance, a person can be young, old or of middle age. However, the limit between these definitions is fluent. A man of 35 years is called young by approximately 75 % people, but 25 % say that he is of middle age. Expressed differently, a person of 35 years is a lot of 'young' and a little bit of 'middle'.

Another example is the fuel consumption of a car. It can be 'low', 'middle' or 'high'. By defining a consumption of 12.0 l/100 km as a the upper threshold of 'middle', it is hard to understand why a car that consumes 12.1 l/100 km should be considered as a car with high consumption.

In fuzzy logic, limits are fluent. The simplest realization is to define a value false (or true) below a threshold and true (or false) above another threshold. In between these limits the degree of truth shifts with a certain function. The example of different temperature ranges explains this functionality (cf. Figure 4.1a). Here, temperatures below 5 °C are declared as 'cold' whereas in the range between 5 °C to 15 °C the term changes from 'cold' to 'warm'. The same is true for the temperature range 20–30 °C, where the definition changes from 'warm' to 'hot'. In the example the term 'warm' applies with approximately 65 % to the temperature $T = 25$ °C. The term 'hot' has a degree of truth of 35 %. So, in a linguistic way, one would say it is 'almost hot'.

For calculations in fuzzy logic the general logic functions need to be adapted to the more complex situation. In general, at least two basic Boolean logical functions are essential to achieve all demanded results [126]. That can be a unifying operator (OR) and an inversion (NOT) or an intersecting operator (AND) and an inversion (NOT). These operators can be combined and transferred to other function, like NAND or NOR. In fuzzy logic, there are analogies to those basic operations. The Boolean AND operator is true if both inputs are also true, for all other cases it is false. Similarly, in a fuzzy system the corresponding function is realized by the minimum of both inputs (cf. Figure 4.1b). The output of the AND function is true if both inputs have a truth degree of '1' and it is false if one input is '0'. However, for values that differ from '1' and '0' the output is the lowest input value. So, if one input has the value 0.9 (very true) and the other input is 0.3 (rather false), the total output is also 0.3 (rather false). Likewise, the analog to an OR function is the maximum of both inputs (cf. Figure 4.1c) and the NOT function can be realized by building the inversion of the input (cf. Figure 4.1d).

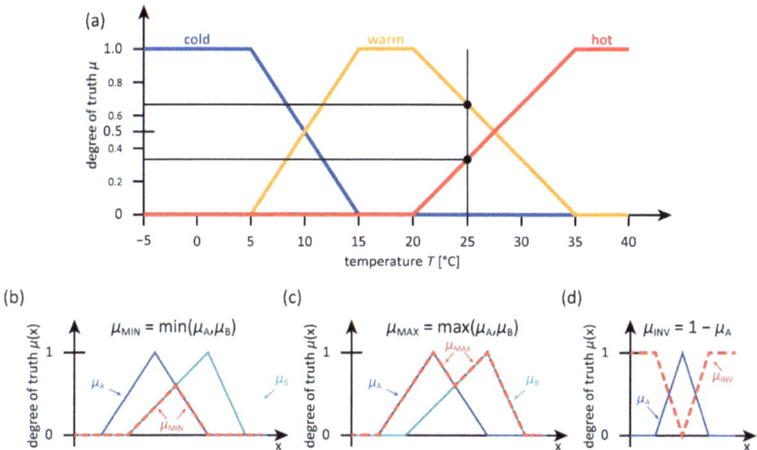

Figure 4.1: Basic principles of Fuzzy Logic – (a) the terms 'cold', 'warm', and 'hot' do not have a strict definition of a certain temperature range. In fuzzy systems this is solved by defining a function for each term that refers a certain degree of truth to a temperature. Accordingly, here the temperature 25 °C is about 65 % 'warm' and 35 % hot. For each temperature the definition is done. (b) Example for a MIN function, the analog to a Boolean AND function. (c) Example for a MAX function, the analog to a Boolean OR function. (d) Example for a INV function, the analog to a Boolean NOT function.

4.2   ReRAM-based Comparator

In 2011 Klimo and Such showed how ideal resistive switches could be used to compute the minimum (MIN) or maximum (MAX) of two applied potentials [122] for implementation of Zadeh fuzzy elementary logical functions. The basic functionality of MIN and MAX gates is the classification of two analogue signals. Each has two input terminals and one output terminal. The gate outputs the lower (MIN) or higher (MAX) voltage of the two input potentials without modification in the predefined operating range. There is a conspicuous similarity of the here demonstrated MIN/MAX gates to the later published AND/OR [127] functionality which is compatible with standard CMOS technology. The reason is that the Boolean logic is a subset of Zadeh fuzzy logic reduced into two value set. These gates could be used in analogue signal processing and could help to realize small-size sorting networks [120] taking some limiting properties into account (Section 4.3 and [128]). The proposed gate structure consists of two anti-serially connected resistive switches.

Like described earlier, two anti-serially connected resistive switches have so far been studied as a possible solution to sneak paths in crossbar arrays of resistive switches [27]. Unlike for the here presented gates, in those complementary resistive switches the electrode in the middle is not tapped [123, 129].

The principle behind the minimum and maximum gates using resistive switches is that when one of the two cells is in a LRS and the other in a HRS a voltage between the inputs will drop mostly across the high resistive cell. If the R_{off}/R_{on} ratio is high enough, it can be assumed that the voltage drop across the cell that is LRS is so low, that the potential in between the two cells (V_{out}) equals the input to the low resistive cell.

The difference between the minimum (MIN) and the maximum (MAX) gate is the assembly of the cells (cf. Figure 4.2a,b). In the maximum gate (b) the active electrodes connect the inputs, leading to the higher potential being passed on to the output. Conversely, for the minimum gate (a) the active electrodes face the output, so that the higher potential applied to the inputs will be blocked.

To verify the theoretical functionality and proof the feasibility of such MIN and MAX gates, microstructure test devices have been fabricated and the electrical behavior was investigated for different cases.

4.2.1 Devices

Two different setups are required to fabricate a MIN and a MAX gate. Just like the original suggested CRS assembly [27] the devices can be built in a single stack. That is also the most area saving approach. However, in terms of area demand, the wiring of three signals can be a limiting property for real implementation. Here, the material system Ta/Ta$_2$O$_5$/Pt, based on the valence change mechanism, was chosen for the RRAM core elements. This system has already proven to show very good CRS results [74].

Figure 4.2 gives an overview of the full setup. The stack consists of two ReRAM elements, one on top (top element) and a second one in reversed polarity underneath (bottom element). For the MIN gate the active electrode is tapped to the middle electrode, so both elements share the active Tantalum layer (cf. Figure 4.2a). The basic setup of the MAX gate is depicted in (cf. Figure 4.2b) with the stack Ta/Ta$_2$O$_5$/Pt/Ta$_2$O$_5$/Ta, where the Platinum layer builds the shared middle electrode. For better contacting properties and prevention of Tantalum oxidation, the whole stack is framed by additional Platinum layers on the bottom and on the top. Such a Platinum framing could be also applied for contacting the Tantalum middle electrode of the MIN gate.

As the stack is fabricated symmetrically, the *I-V* characteristic has no polarity dependency. Furthermore, a differentiation between MIN and MAX gate assembly simply by the *I-V* curve is not possible. However, for the same switching polarity the MIN gate is internally switched to a diametrically opposed state compared to the MAX gate. For example, for a positive voltage exceeding the threshold $V_{th,2}$, the configuration LRS/HRS (top to bottom) results for the MIN gate, but HRS/LRS results for the MAX gate. The inverse configuration for a negative switching voltage below the threshold $V_{th,4}$ are obtained.

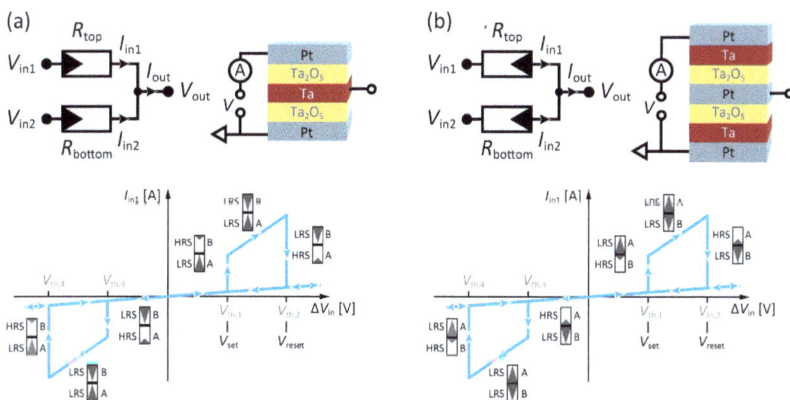

Figure 4.2: MIN and MAX gate principle – (a) the assembly of two resistive switches for a Minimum gate is realized by the Pt/Ta$_2$O$_5$/Ta/Ta$_2$O$_5$/Pt-stack, (b) while the maximum gate is built with the reversed CRS stack, namely Pt/Ta/Ta$_2$O$_5$/Pt/Ta$_2$O$_5$/Ta/Pt. The corresponding *I-V* characteristics exhibit the same behavior. As the CRS stack basically just reversed, the stack for the MIN-gate is declared as BA (from top to bottom) and the MAX-gate is declared as AB. The threshold voltages $V_{th,1}$ and $V_{th,3}$ specify the set point V_{set} of the individual resistive element, while $V_{th,2}$ and $V_{th,4}$ define the reset voltage V_{reset}.

This intrinsic difference in state is the key property for the MIN/MAX-gate functionality. Three essential conditions need to be fulfilled to operate proper fuzzy logic:

- The device must enable reversible toggling between two states. That is the case for CRS devices.
- One of the cell elements needs to be in HRS whereas the other element is in LRS. Being in the 'ON'-window leads to unintended behavior.
- The resistance of the high resistive element needs to be sufficiently higher than the resistance of the low resistive element – in other words, the R_{off}/R_{on} ratio needs to be adequately high.

Especially, the second point leads to an important constraint. It means that for proper operations the threshold voltage $V_{th,2}$, respectively $V_{th,4}$, needs to be exceeded. Accordingly, the applied voltage difference $V_{in1} - V_{in2} = \Delta V_{in}$ has a minimum voltage level, which is defined by the threshold voltage of the CRS cell. So, material systems with comparable low switching voltages are beneficial for a more accurate operating range with low offset. However, these material systems often do not allow very high maximum voltages V_{max} without cell destruction. A sophisticated tradeoff between both limits is required.

The third point, the R_{off}/R_{on} ratio, is an indicator for the quality of the output result. The two input voltages V_{in1} and V_{in2} cause the output voltage V_{out} by the voltage divider which is given by the resistances of the resistive elements R_{top} and R_{bottom}. Accordingly the following equation holds for both gate setups:

$$V_{out} = \frac{R_{top}}{R_{top} + R_{bottom}} \cdot V_{in2} + \frac{R_{bottom}}{R_{top} + R_{bottom}} \cdot V_{in1} \qquad (4.1)$$

For $R_{top} \gg R_{bottom}$ Equation (4.1) simplifies to $V_{out} \approx V_{in2}$ and for $R_{top} \ll R_{bottom}$ it simplifies to $V_{out} \approx V_{in1}$, respectively. Thus, a high R_{off}/R_{on} ratio is favorable since it decreases the transition offset, i.e. the difference between input and output voltage.

4.2.2 Mechanism

By this consideration, the mechanism of the gate also gets clear. For a positive ΔV_{in} higher than the threshold $V_{th,2}$ the MIN gate switches to $R_{top} = R_{off}$ and $R_{bottom} = R_{on}$, so HRS/LRS. Consequently, $R_{top} \gg R_{bottom}$ is the case and the gate redirects V_{in2} which has to be the lower input voltage when ΔV_{in} is positive. Correspondingly, the MAX gate is switched to LRS/HRS for positive ΔV_{in}. Since $R_{top} \ll R_{bottom}$, the MAX gates transfers

V_{in1} which is the higher input voltage. For negative input voltage differences lower than $V_{th,4}$ the gates redirect in each case the alternate input voltage.

The effect can be physically explained by simplification of the switching states. By assuming that an HRS is an infinitely high resistance and an LRS is infinitely low, the HRS becomes a perfect insulator and the LRS becomes a perfectly conducting connection. Now, it is obvious that the combination HRS/LRS connects the second terminal (V_{in2}) to the output, as well as the combination LRS/HRS redirects the first terminal (V_{in1}).

Table 4.1 gives an overview of the possible transition states. In all considerations it is assumed that the gate has no crucial electrical load, i.e. the output current I_{out} is negligibly low.

	Input			CRS state		Output	
#	V_{in1}	V_{in2}	ΔV_{in}	**MIN**	**MAX**	**MIN**	**MAX**
1	Low	Low	$< V_{th,1}$ $> V_{th,3}$	LRS B / HRS A	HRS A / LRS B	V_{in1}	V_{in2}
	Low	Low	$< V_{th,1}$ $> V_{th,3}$	HRS B / LRS A	LRS A / HRS B	V_{in2}	V_{in1}
2	High	Low	$> V_{th,2}$	? B / ? A $\rightarrow$ HRS B / LRS A	? A / ? B $\rightarrow$ LRS A / HRS B	$? \rightarrow V_{in2}$	$? \rightarrow V_{in1}$
3	Low	High	$< V_{th,4}$	? B / ? A $\rightarrow$ LRS B / HRS A	? A / ? B $\rightarrow$ HRS A / LRS B	$? \rightarrow V_{in1}$	$? \rightarrow V_{in2}$
4	High	High	$< V_{th,1}$ $> V_{th,3}$	LRS B / HRS A	HRS A / LRS B	V_{in1}	V_{in2}
	High	High	$< V_{th,1}$ $> V_{th,3}$	HRS B / LRS A	LRS A / HRS B	V_{in2}	V_{in1}

Table 4.1: Truth table for MIN/MAX gates – It is assumed that $R_{off} \gg R_{on}$ and the input voltages do not exceed the working range of the ReRAM elements.

4.2.3 Experimental

To verify, the previously explained behavior, measurements on the fabricated Ta_2O_5 CRS stacks were performed. To this end, an electrical characterization setup consisting of a Keithley 4200-SCS and an Agilent B1500A was used. The test devices were fabricated as micrometer crossbars and structured by UV-photolithography, chemical and physical dry etching on thermally oxidized p-type Silicon wafers. 5 nm Titanium (Ti) and 30 nm Platium (Pt) are deposited by magnetron sputtering and build the lowest layer, which also serve as the bottom electrode. For the MIN-gate a 10 nm Tantalum oxide (Ta_2O_5) and 10 nm Tantalum (Ta) build the middle electrode. The top electrode is structured by 10 nm Ta_2O_5 and 25 nm Pt. The result is the material stack like in Figure 4.2a. These samples were fabricated within the work of T. Breuer [77, 130].

For the MAX-gate, an alternative planar CRS assembly was used, which connects two single ReRAM elements to an anti-serially setup. The fabrication of planar CRS devices is less elaborate since both elements are fabricated simultaneously. Firstly, a long electrode is formed by the Ti and Pt layer on the p-type Silicon substrate. That electrode serves as the middle electrode. Then, 10 nm Ta_2O_5, 10 nm Ta and 25 nm Pt are deposited and structured to two electrodes. One of them is used as the top electrode, the other one as the bottom electrode (cf. Figure 4.2b). The experiments were conducted within [130].

4.2.4 Characterization

The fabricated test samples have been studied by electrical characterization. First, the single ReRAM elements have been separately investigated. By applying a voltage to the middle electrode and grounding the other electrodes, electroforming and bipolar switching were performed. The initial resistance of the virgin elements was in the range of several hundred $k\Omega$ to a few $M\Omega$. The electroforming took place at around 1.7 V [130], whereas a current compliance (CC) of 500 μA was used. The CC limits the conductivity of the cell. A lower CC decreases the maximal operation current in the single bipolar switching cell and in the final CRS device [74, 130]. Voltage cycling reveals the characteristic properties of the resistive switches. In the VCM-based ReRAMs the RESET does not occur abruptly but gradually, whereas the SET event can be observed at a sharp switching voltage. The gradual RESET begins at a voltage $V_{reset} \approx -0.6$ V and the abrupt SET at $V_{reset} \approx 1.0$ V. Since VCM elements also do not have a linear ohmic behavior below the switching voltages, the resistance states need to be defined for a certain voltage or voltage range. Here, a read voltage range of ±0.5 V points out that the LRS has a resistance of $R_{on} \approx 1\,k\Omega$ and the HRS can be measured with values of $R_{off} \approx 100\,k\Omega \dots 1\,M\Omega$. Thus, the devices have a resistance ratio of about $R_{off}/R_{on} \approx 100\dots1000$, which is sufficiently high for the MIN/MAX-gate operation (cf. Section 4.2.1). Furthermore, endurance test have shown that no significant device degradation appeared for more than 10^6 switching cycles.

With these results, the full CRS stack shows an *I-V* behavior like in Figure 4.3a. Despite the slight variation in the fabrication, both gate types are equivalent in terms of CRS switching. Figure 4.3b shows a scheme for pulse driven experiments with transient current measurements. This pulse approach is also the interesting mode for real MIN/MAX applications. CRS switching can be observed by the current peaks, that are caused by the instantaneous voltage stimulus.

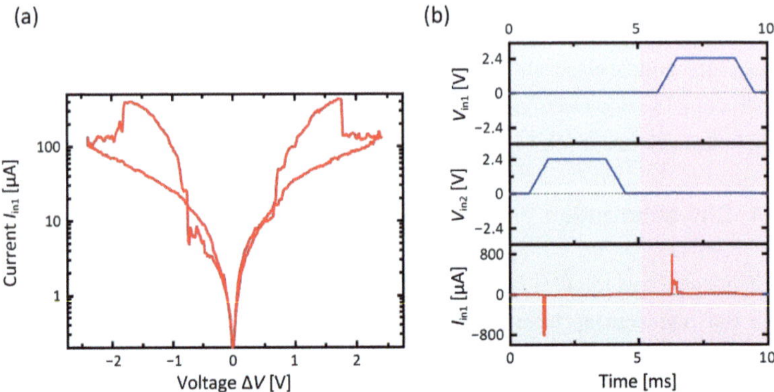

Figure 4.3: Ta$_2$O$_5$-based CRS cells – both devices, MIN and MAX gate, reveal a similar *I-V* curve. The cell switches in the LRS/LRS state around 1 V. A full RESET can be ensured for voltages around 2.0–2.4 V. Since the electroforming was performed with a CC of 500 µA, the maximum current also does not exceed this value. Adapted from [130].

4.2.5 MIN-Gate

The characteristic effects of the MIN-gate function are summarized in Figure 4.4. Three different control modes are demonstrated to illustrate the typical behavior of the experimental implementation. Different voltage levels were chosen to present the variability of the gate. All measurements are plotted as a function of time since this agrees with the final application mode. On top, small icons of the CRS cell indicate the current logic state and shall support the understanding of the particular output curve. The first two signals are the applied input signals. Below, the measured device current and output voltage are shown. First, the CRS device is set to the initial state by applying a high voltage for V_{in1} and a low voltage for V_{in2}. This is not shown in the Figure.

In all three cases the signal combinations equivalent to the signal sequence, which are indicated by the first column in Table 4.1, are applied. In general, the sequence does not influence the result of the gate. But the previous state can affect the dynamics of the observed output signal V_{out}.

In Figure 4.4a, a quasi-static voltage ramp is applied to each input terminals. In this case a positive voltage ramp up to 1.2 V is defined as a high signal ('H') and a negative voltage ramp down to −1.2 V is a low signal ('L'). As also indicated by the CRS icons, only for the signal combination V_{in1} = 'L' and V_{in2} = 'H' switching occurs. This event can be observed in the current measurement by the negative current peak. In the output measurement V_{out}, the switching causes a deviation from the triangular form of the applied signals. First the ramp rises like the signal V_{in2}, but when the total voltage ΔV_{in} exceeds the SET or RESET voltage the gate switches to the correct state. Since in this case the measurement is performed in quasi-static mode, a gradual switching can be observed where the CRS device has the LRS/LRS state for a certain time.

Figure 4.4b shows a static voltage experiment. Here, no transient events are detectable since the experimental equipment for this test does not enable an adequate time resolution. Consequently, no current curve was depicted. A low signal corresponds to 0.6 V and a high one to 3.0 V. For all four signal combinations the output reveals the correct final logic result.

To come closer to a real application, pulse measurements were performed (cf. Figure 4.4c). Now, the convention is that a low signal corresponds to 3.0 V and a high signal is 5.5 V. Also for these values, the gate outputs deliver the true logic results as seen in the output voltage V_{out}. Similar to the quasi-static mode (cf. Figure 4.4a), the CRS stack switches for V_{in1} = 'L' and V_{in2} = 'H'. However, only a very short current pulse is detected. The LRS/LRS state can be observed for very short time in the output voltage V_{out}.

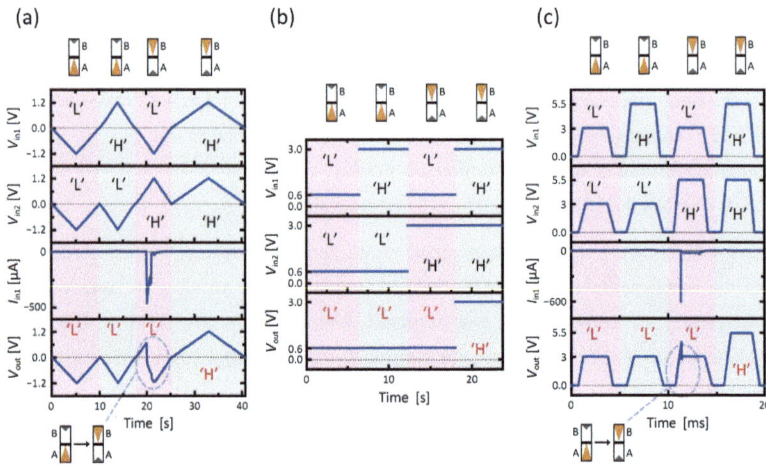

Figure 4.4: Experiments on MIN gate – The MIN operation is implemented by three different voltage modes: quasi-static voltage sweep (a), base voltage (b) and voltage pulse (c). The resistance scheme at the top indicates the final CRS state. If a change of the CRS state is observable in the measurement, the switching is illustrated explicitly by the resistance scheme at the bottom. The graphs show from top to bottom: voltage signal lines at V_{in1} and V_{in2}, current signal line (only a, c) and the detected voltage signal at V_{out}. (b) does not include the current signal line, since no switching dynamics are detected. Adapted from [130].

4.2.6 MAX-Gate

Similar to the MIN gate also three voltage schemes were applied to the MAX-gate. Figure 4.5 shows the performed measurements with (a) voltage pulses and (b, c) two different quasi-static voltage sweeps. Here, no scheme with a constant base voltage is presented but two different quasi static modes. Thereby the influence of different signal sequences is demonstrated. In all three plots, the two applied input voltages V_{in1} and V_{in2}, the measured input current I_{in1}, and the measured output voltage V_{out} are illustrated. Like for the MIN-gate, all cells are initially switched by V_{in1} = 'H' and V_{in2} = 'L' except for the third (c) measurement. There the gate was initialized by the reversed polarity V_{in1} = 'L' and V_{in2} = 'H'. The icons on top of the graphs show the according CRS state of the MAX gate.

The first two signal sequences are equivalent to the signal sequence of Table 4.1. The third one is modified to show that the sequence does not influence the functionality.

Figure 4.5a shows a pulse measurement with the convention 'H' = 3.0 V and 'L' = 1.1 V. The switching takes place when V_{in1} = 'L' and V_{in2} = 'H' are applied. Theoretically the short switching event is identifiable in the output voltage. For this case, however, the irregularity in the rising ramp is very hard to recognize. Here, the switching can be easier detected in the current regime. Besides the static current through the high resistive cell, the current peak is observable. The gate redirects correctly the maximal input signal to the output in all four cases.

In Figure 4.5b,c the high signal is set to 4.7 V whereas the low signal is set to 2.7 V. The measurement is performed with quasi-static triangular voltage ramps. Thus, both measurement were conducted with comparable conditions. But whereas the sequence of (b) corresponds to Table 4.1, in (c) the last two signal combinations are interchanged. Furthermore, the initialization voltage is changed. Thus, two switching events take place – the first one when the signals V_{in1} = 'H' and V_{in2} = 'L' in phase 2 are applied and a second one when V_{in1} = 'L' and V_{in2} = 'H' are applied in the last phase. It can be seen that the functionality of the sorting behavior is not affected by altering pre-conditions and the order of applied signals. In all cases, the output signal has the correct logic state.

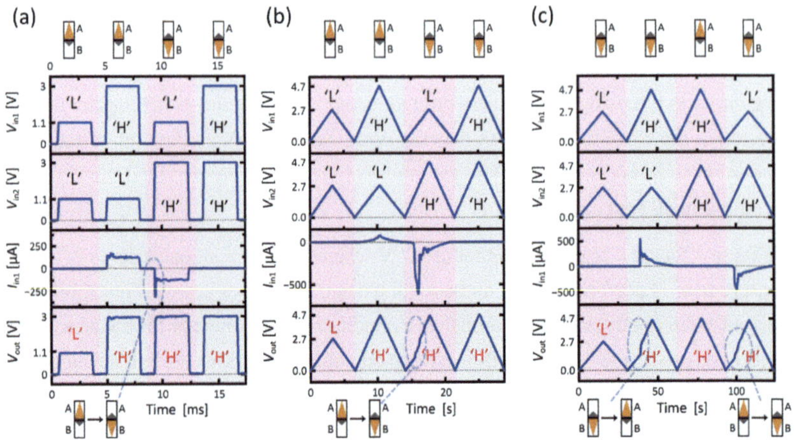

Figure 4.5: Experiments on MAX gate – The MAX gate function is implemented by different conditions: voltage pulse (a), quasi-static voltage (b, c). Additionally, (b) and (c) distinguish from each other by the reversed initialization voltage polarity and by a modified sequence for V_{in1} and V_{in2}. The resistance scheme at the top indicates the final CRS state. If a change of the CRS state is observable in the measurement, the switching is illustrated explicitly by the resistance scheme at the bottom. The graphs show from top to bottom: voltage signal lines for V_{in1} and V_{in2}, current signal line and the detected voltage signal at V_{out}. Adapted from [130].

4.2.7 Disscussion

By these experiments it could be shown that the fabricated gates fulfil the predicted logic function. A certain switching time needs to be taken into account. This time depends mainly on the voltage difference ΔV_{in} or slew rate. A higher voltage leads to faster operations. In the conducted experiments, the utilized impedance converter with bandwidth of 8 MHz and a slew rate of 2.8 V/µs limits the resolution. In general, it can be expected that the switching speed is not a limiting property of ReRAM since switching speeds below 200 ps have been demonstrated [131]. In terms of application, the bigger challenge is to optimize the sense amplifier for the output signal. The demanded properties for this sensing circuit are high input resistance, wide bandwidth and very high slew rates. However, conventional impedance converters do not combine all these requirements. Especially, integrated circuitries have typically poorer input impedances. In general, the input impedance of elements of the next logic stage has an impact on the current stage, i.e. the loading of the output voltage signal has to be considered for circuit design. Many intended applications for MIN/MAX gate utilization, like audio signal processing for example, have relaxed requirements as the operations timing is in the range of milliseconds. Here, the standard circuits would be quite sufficient.

For the input values V_{in1} and V_{in2} some limitations need to be taken into account. The theoretical considerations allow voltages in arbitrary ranges as long as the differential voltage ΔV_{in} exceeds the threshold voltages $V_{th,2}$ or $V_{th,4}$, respectively. If not, the voltage divider of the CRS cell would cause false behavior regarding the logic function. Furthermore, the level of the input voltage difference ΔV_{in} needs to be considered, since the total voltage drop has an impact on the switching kinetics, which is primarily affected by the non-linear voltage to time characteristic of ReRAMs [64, 132-134]. This is especially of interest for the pulse driven mode since it needs to be ensured that the voltage drop his high enough to switch the CRS cell within the given pulse width. Improvements of the gate could be achieved by material engineering of the ReRAM device stack (reducing the thresholds $V_{th,2}$ or $V_{th,4}$) or optimizing the pulse parameters, e.g. rising and falling time.

Compared to conventional CMOS approaches, the presented ReRAM approach offers smaller unit array and superior scaling properties. A basic requirement to keep the power consumption low is to use ReRAM devices offering large high resistive states (HRS) and fast switching from LRS/HRS to HRS/LRS. In general, further improvements of ReRAM cell performance in terms of reliability, cycle-to-cycle variance and endurance are required to enable ReRAM based memory and logic applications. Although ReRAM devices enable energy-efficient operations in principle [71], the question whether the energy-efficiency of ReRAM-type MIN/MAX gate-based circuits is better than comparable CMOS circuit cannot be answered without knowing the area of application (e.g. sorting or audio signal processing) and the actual circuit implementation.

As shown in the following Section 4.3, the presented MIN and MAX logic gates offer limited performance in terms of interconnectivity. Since there is no active signal restoration within the passive memristive device, additional amplification circuits between the stages would be needed. One possibility is to add analog buffer stages in the circuitry, which recover the redirected signal. The high input and low output impedance of those stages would also have a positive effect on the system. Another application can be found in replacements of clocked pass-gate transistors for cascading structures like in FPGAs [135]. However, the system can work adequately enough if assembled in small circuit blocks without buffer. For example, implementation of area and energy efficient memristive sorting networks are feasible and can be beneficial compared to area demanding CMOS circuits.

4.2.8 Conclusion

As an example of two fundamental fuzzy logic devices, the MIN and the MAX gate functionality was shown and analyzed for the example of Ta_2O_5-based resistive switches. The input voltage difference turned out to be one important factor in the functionality, since it the settling time of real resistive switches depends on the applied voltage and influences the switching speed of the whole gate. MIN and MAX gates can be simply derived from technologically established CRS cells by adding an extra terminal to the middle electrode. The feasibility of this device setup enables interesting analogous singnal processing applications and is the fundament for sorting network which will be described in the following Chapter.

4.3 ReRAM-based Sorting Network

The previously described feasibility of minimum and maximum gates enables the realization of an analog comparator. This comparator is the basis for the sorting network, which will be discussed in the following section.

4.3.1 System of Sorting Networks

Sorting networks can realize sorting tasks directly in hardware. The key component is a comparator, which can be efficiently implemented using resistive switches [122, 123].

The comparator, the central element of sorting networks, compares two inputs and returns the lower input value on one output and the higher input value on the other output. Thus, the functionality of a minimum (cf. Figure 4.6a) and maximum gate (cf. Figure 4.6b), which were investigated in the section before, is needed in one device. In fact, combining a memristive MIN and MAX gate in parallel, leads to a functional comparator depicted in Figure 4.6c.

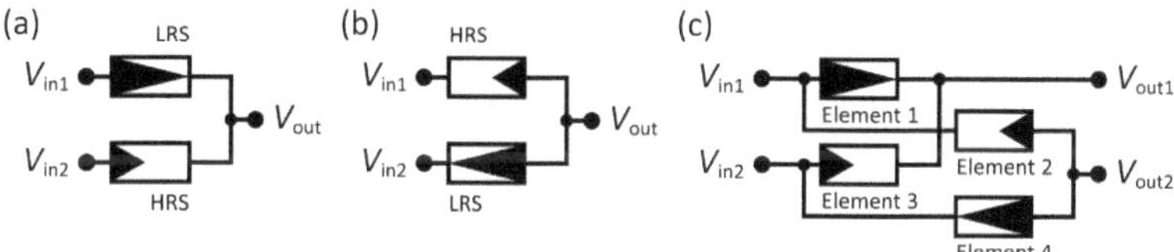

Figure 4.6: Comparator – (a) a minimum gate (MIN), (b) a maximum gate (MAX), both with one high resistive (HRS) and one low resistive (LRS) cell. (c) a comparator consisting of one MIN and one MAX gate.

For this comparator, the output V_{out1} returns the minimum of the input potentials V_{in1} and V_{in2}, while output V_{out2} returns the maximum. Depending on the inputs either Element 1 and 4 are low resistive while Element 2 and 3 are high resistive (as in Figure 4.6) or if Element 2 and 3 are switched to the LRS, Element 1 and 4 are switched to the HRS. This way the inputs can be sorted. Of course, the structure requires stable input potentials. By interconnecting comparators, sorting networks are built.

The basic comparator structure (cf. Figure 4.6c) can be directly implemented as 2×2 crossbar array [123]. However, the complete sorting network requires more complex wiring than provided by a basic crossbar array. Thus, the normalized mean area footprint of such a sorting network is larger than $4 \cdot F^2$.

For a real implementation, electrochemical metallization cells are well suited since they do not only offer high R_{off}/R_{on} but also low threshold voltages. As explained in [123] large threshold voltages (due to highly nonlinear switching kinetics) lead to permutation networks instead of sorting networks.

A four input network implementing Batcher's odd-even merge-sort [136] was simulated (see Figure 4.7). Here, the comparators sort the inputs in ascending order. According to that, the networks outputs order is also ascending ($V_{out1} < V_{out2} < V_{out3} < V_{out4}$). The idea behind the odd-even merge-sort is that first ordered sequences (of two elements) are created, then the odd lines are merged as well as the even ones and at the end the output of the odd merge is compared to that of the even merge.

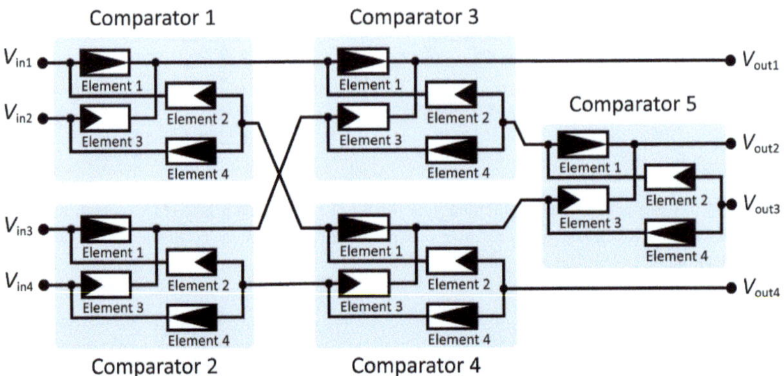

Figure 4.7: Batcher's sorting network

Criteria for the choice of a sorting network are the number of comparators that are needed and the caused delay time. The time needed by a network to sort is proportional to the number of comparators in the longest path. The simulated network uses 5 comparators arranged in 3 stages. This is the optimum for a four-input network [137]. For other situations, where another number of inputs is demanded, another sorting network structure can be more efficient and accordingly the ideal implementation is not necessarily the Batcher's odd-even merge sort.

4.3.2 Simulation Models

Simulations were performed with Cadence Spectre. For the simulations a model of silver iodide (AgI) ECM cells by Menzel et al. described in [138] is used. The model is built on measured data and has been proven to be reliable [139]. It is a one-dimensional memristive model, which can be represented by the electric circuit in Figure 4.8. The change of the gap (state variable x) between the filament and the active electrode is modeled by Faraday's law. The tunneling current is described according to Simmons, the electron transfer process is described by the Butler-Volmer equation and the ion transport by the Mott-Gurney law. Nucleation is not considered in the model, but according to the measurement data switching at very low voltages is very slow. A lower boundary was inserted in the dynamical model, so that for voltages lower than 10^{-2} V no switching will occur. The actual parameters are given in Table 4.2.

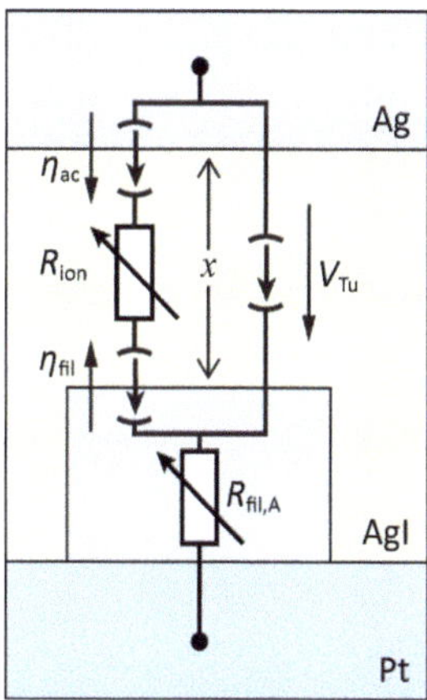

Figure 4.8: Simulation Model – Equivalent circuit diagram for the AgI ECM model. The cell consists of a three layer stack: Pt/AgI/Ag. The gap x between the conductive filament ($R_{fil,A}$) and the active Cu electrode is the state variable which is altered by the ionic current. The ionic current (left branch) is defined by the Butler-Volmer equation (overpotential η_{fil} and η_{ac}) and the ionic hopping resistance R_{ion} is described by the Mott-Gurney law. The electronic tunneling current (V_{Tu}) is defined by the Simmons tunneling equation.

Parameter	Description
Resistance of the electrodes	76.4 mΩ
Switching layer thickness	20 nm
Mass density	10.49 g·cm^{-3}
Filament area	12.57 nm^2
Molecular mass	1.79·10^{-22} g
Conductivity of the active filament/electrode	5.88·10^7 S·m^{-1}
Barrier Height	3.36 eV
Effective electron tunneling mass	0.023 × 9.1·10^{-31} kg
Temperature	300 K
Charge transfer coefficient	0.3
Ionic charge of the cations	1
Exchange current density	3.2·10^5 A·m^{-2}

Table 4.2: Simulation Parameters of an AgI ECM cell.

AgI cells strongly exhibit the earlier established criteria of a low threshold voltage and high R_{off}/R_{on} ratio. Due to the low threshold, a sorting network which implements AgI cells is even for small input differences suitable. Note that for this kind of application the relative short-term retention of AgI cells is no drawback as it would be for storage applications.

First, proof of principle simulations was performed with a simpler model, which offers predefined threshold voltages. The threshold devices switch between defined R_{on} (5 kΩ) and R_{off} (1 MΩ) as soon as the applied voltage is greater/smaller than a programmed set/reset voltage (±1 V). So, unlike the physical ECM cells, the switching of the threshold devices does not depend on the time the voltage is applied. These devices implement a basic hysteresis function and only have two clearly defined ohmic states.

4.3.3 Sorting Network – Circuit Implementation Constraints

First simulations were done with the threshold model. The network was tested for a wide range of input sequences.

Unfortunately, depending on the input sequence and initial device states unintended switching may occur. If the cells are initially low resistive or switch to a low resistive state during sorting, sneak path currents can flow through low resistive cells to the inputs. This back propagation can lead to unwanted resets. An example of such an unwanted feedback loop is depicted in Figure 4.9. As switching occurs in Comparator 3, four low resistive cells connect inputs 1 and 3. Instead of Element 3 in Comparator 3 Element 1 in Comparator 2 resets.

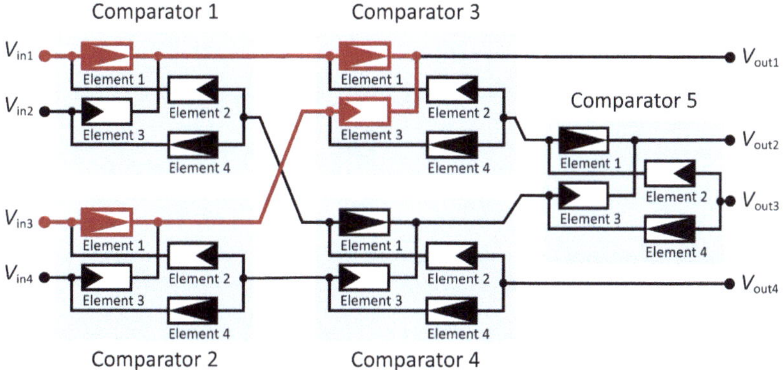

Figure 4.9: Feedback loop in sorting network – Example of a feedback loop leading to unwanted resets in arbitrarily configured networks.

Basically, the sorting should propagate continuously from the input stage to the output stage. The settled devices of Comparator 1 and 2 should not be altered anymore when Comparator 3 and 4 are being settled. Unfortunately, due to the passive nature of the network, voltages large enough to switch a device unintentionally may also occur at previously settled devices caused by the presence of a voltage divider effect. Note that there are some scenarios in which no such problems occur, as was shown in [120].

This can happen for inputs like $V_{in1} = 3$ V, $V_{in2} = 6$ V, $V_{in3} = 9$ V and $V_{in4} = 12$ V if Elements 2 and 3 are initially in LRS for all the comparators. This is just one example, there are other unwanted feedback loops. To prevent these feedback loops it seems necessary for the ECM cells to be initially in the HRS.

Configuring all elements to HRS is the first requirement. However, even if all elements are initially in the HRS, the network does not sort the inputs correctly for all possible initial setups. Because of that, for each stage transmission gates were introduced into the network in between the stages (see Figure 4.10). By this topology, each stage sets separately and correct functionality was achieved. To implement these transmission gates in the simulations, a simplified transistor switch model is used. The switches are clock controlled, so that after setting of the first stage, switches in switch-network 1 close and later on, after Comparator 3 and 4 are set, switches in switch-network 2 close. This network can sort inputs in ascending order.

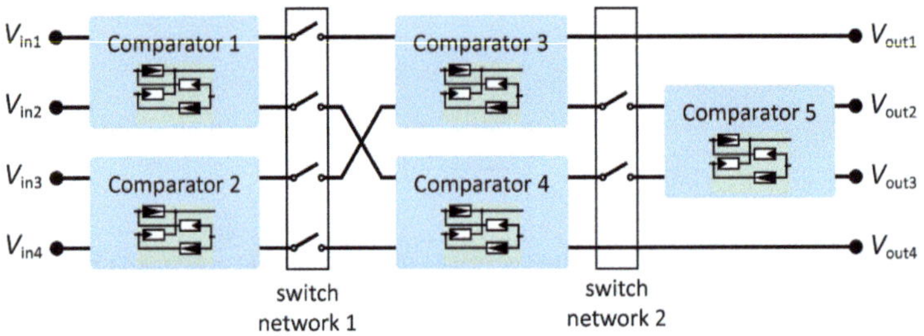

Figure 4.10: Sorting network with switches in between the stages – The switch networks 1 and 2 are clock controlled, so that each stage can set individually, leading to a more stable system.

4.3.4 Derivation of the Allowed Input Range

The input voltages may not have arbitrary values but should be selected carefully to avoid a malfunctioning sorting network. In the following, the allowed input voltage range for a specific sorting task can be derived. First, ΔV as the largest difference between two arbitrary input voltages of V_{in1} to V_{in4} is defined. Moreover, an equidistant spacing between the inputs is assumed, i.e. 0 V, ⅓ V_{max}, ⅔ V_{max} and V_{max}. In the following it is shown that there is an upper and a lower limit for ΔV. The upper limit is related to the prevention of unintended RESETs, much like those previously described, while the lower limit results from the need for defined SET operations. Moreover, it will be shown that the limits also depend on the R_{off}/R_{on} ratio.

The lower limit is a little more than twice the SET voltage. This property can be directly deduced from the basic comparator circuit where two devices in HRS are connected in series. Due to the voltage divider property, only half of the applied voltage is present at the device that is intended to switch to LRS, i.e. the lower voltage limit is a little more than twice the SET voltage.

The upper limit is a function of the RESET voltage. If the voltage differences between the inputs are higher than the upper limit, unwanted resets occur in Comparator 1 and 2. For such a case, the network has been decomposed by star-delta transformations for each voltage source. The total voltage drop across one of the cells that resets in the simulation has been superimposed assuming equal differences between voltages in descending order. If the superimposed voltage equals the reset voltage, the difference between the input potentials is at the upper limit. The results of the calculation are in accordance with the simulations. In Figure 4.11 the maximum difference between the inputs, for which no unintended RESETs occur, is plotted as a function of the reset voltage and R_{off}/R_{on}. The upper limit depends linearly on the reset voltage.

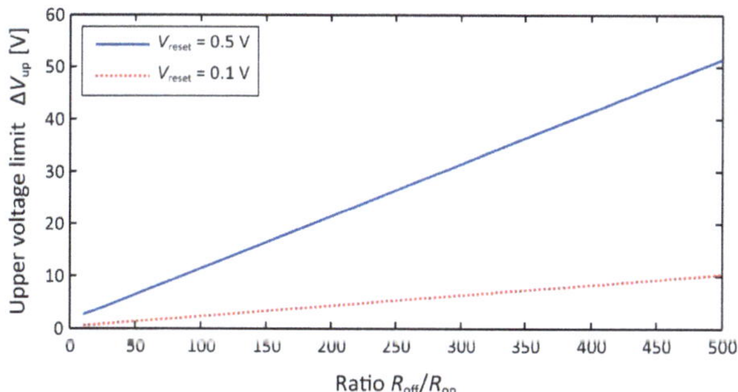

Figure 4.11: Upper limit for the voltage difference between the inputs for which correct sorting is possible with the threshold model as a function of R_{off}/R_{on} and the reset voltage for the case of equidistant input voltages in descending order.

Note that the restrictions for the input voltages only have a minor impact on the feasible network size since only the number of low resistive devices in a path is increased in larger networks. As long as R_{off}/R_{on} is sufficiently large, no negative impact will be present.

Here, the approximation

$$\Delta V < V_{reset} \cdot \frac{1}{x} \cdot \frac{R_{off}}{R_{on}} \tag{4.2}$$

should hold for the given four-input network, while x depends on the specific devices. Here, $x = 5$ holds.

4.3.5 Memristive ECM Model Simulations

After finding the network topology and input value constraints for the simplified model, simulations using the memristive ECM model were conducted. As before switches separated the stages and all elements were pre-configured to the HRS. As for the simple model, correct sorting was obtained for the dynamical simulation model. Figure 4.12a depicts the input potentials, the potentials applied to the switches (cf. Figure 4.12b) as well as the resulting outputs for a simulation using the AgI cells (cf. Figure 4.12c).

Interestingly, using the AgI-cells, two consecutive sorts will work whereas this was not the case for the simple model using predefined thresholds. That is due to varying resistance values (R_{on}) for the low resistive state. Depending of the cell's history two LRS cells can still differ in its resistance up to one order of magnitude, since LRS and HRS only defines a certain range of resistances (R_{on}) and (R_{off}). The length of the filament (AgI thickness minus gap) of each cell is shown in Figure 4.13. It shows that in Comparator 1 and Comparator 2 the two cells that switch to the LRS are influenced by the second stage and reach two different states. However, since the actual LRS will be affected by variations, the network has to be switched back to HRS after each run for a stable sorting operation.

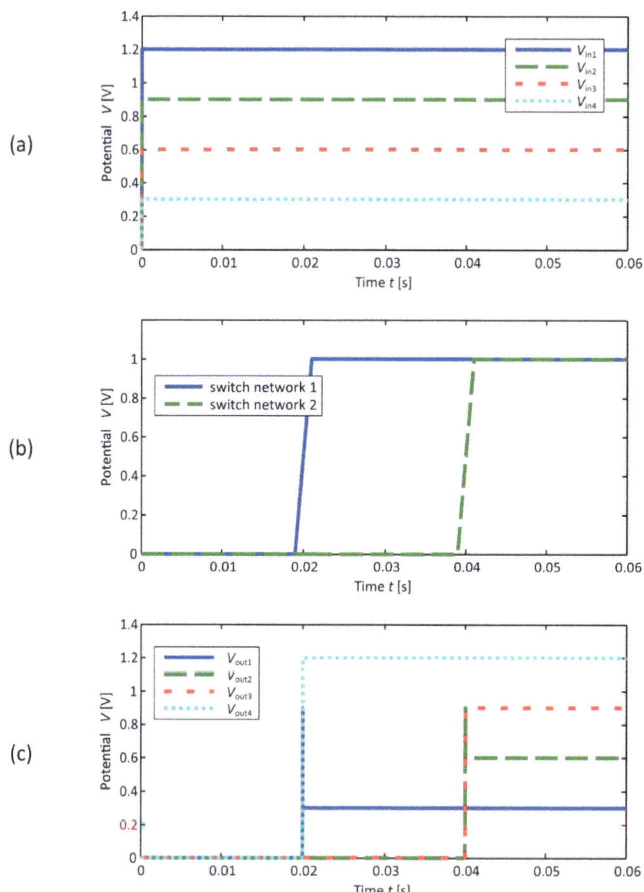

Figure 4.12: Corresponding to Figure 4.10, the potentials at the inputs (a) and at the outputs (c) are depicted for a simulation using the AgI model. The switches in the switch (b) open for voltages higher than 0.5 V. The inputs get sorted correctly.

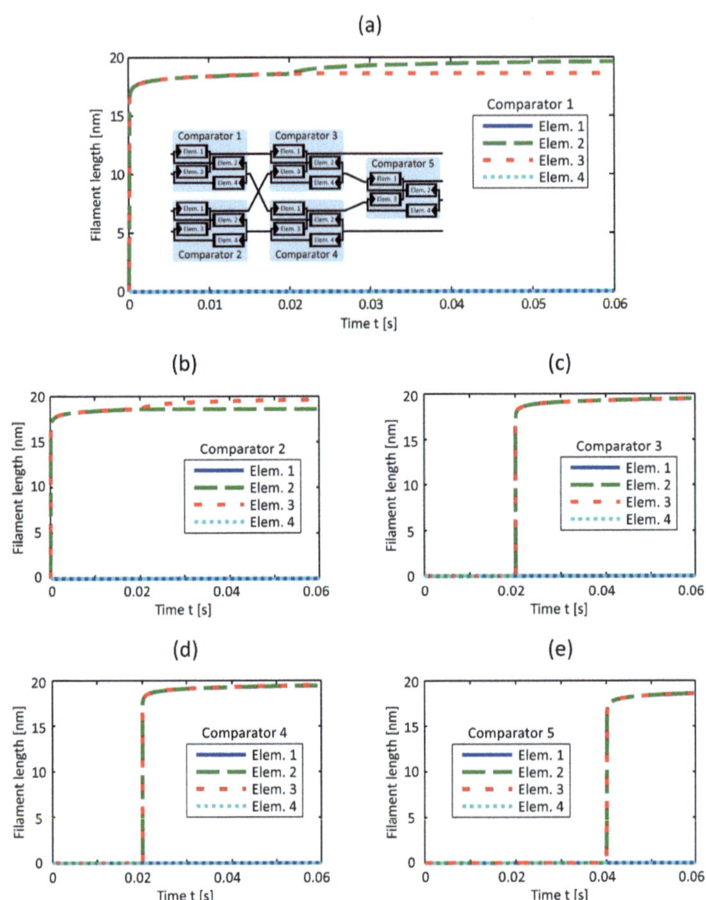

Figure 4.13: Length of the filaments in all the cells for a 20 nm thick AgI layer and potentials as in Figure 4.12. (a) shows the filament length for the four cells of the first comparator. The second and third comparator are depicted in (b) and (c) and the fourth and fifth in (d) and (e).

4.3.6 Discussion

In general, there are the following constraints which must be considered to enable proper operation under any condition:

- All elements of the sorting network must be initialized to the HRS before every sorting operation.
- Stages of the network must be separated by transmission gates, which open stage by stage.
- The input voltage difference at any two inputs of the sorting network must be larger than about two times V_{SET}.
- The input voltage difference at any two inputs of the sorting network must be smaller than a multiple of V_{RESET} (varying with R_{off}/R_{on})

For AgI-type ECM devices the difference between the inputs does not need to be very large, hence this constraint is relaxed. In contrast, too large voltages can become an issue since V_{RESET} is very small for AgI cells. However, for real devices very high voltages would not be recommendable anyhow, as they might lead to a breakdown of cells. Voltages in the range of some millivolts to volts are acceptable. (Small R_{off}/R_{on} ratios could cause also problems in this respect, but for ECM cells this is not an issue since $R_{off}/R_{on} > 1000$ are in general available.) In general, the device selection should care for device's SET voltage, RESET voltage and R_{off}/R_{on} ratio with respect to the intended input potentials to enable a correct sorting. The variability of the SET and RESET voltage and the resistive states may also impact the sorting results. Thus, devices offering low variability should be selected.

The needed time for the cells to switch depends on the applied potentials. The required time of the full sorting network depends on the number of comparators within the longest signal path. Hence, the device properties and input signal range define the clocking of the transmission gates.

4.3.7 Comparison with CMOS

To evaluate the potential and drawbacks of the proposed ReRAM-based sorting network, a comparison with established CMOS techniques is required. However, this comparison reveals a major obstacle. CMOS implementations are typically designed to perform binary information sorting [140-142]. That means that the existing components are not directly applicable on this analog sorting task. The conversion of analog signals by AD- and DA-converters is a logical solution, however, the required overhead for each single input and output channel would be enormous compared to the ReRAM-based solution [143, 144]. That means that the CMOS solution is disadvantageous in terms of area demand and power consumption. Nevertheless, it is likely that a conventional digital implementation can compete in terms of sorting time.

To overcome the need of an AD-/DA-conversion an implementation of an analog voltage comparator could be a solution. However, it must be noted that a single operational amplifier does not meet the requirements, since its output is defined by the supply voltage. A function to forward the inputs voltages would still be necessary. There are a few analog sorting components available. For example, a maximum voltage selector realizes the function of the previously described MAX-gate [145, 146]. An analog voltage comparator can be built together with a corresponding minimum voltage selector. As described in the mentioned references, the amount of additional transistor implementation is very large compared to the proposed fully passive implementation. That is also the case for neural-network inspired approaches [147, 148].

4.3.8 Conclusion

On the example of a four-input odd-even merge-sort network sorting using electrochemical metallization memory cells was shown. It was demonstrated, that AgI cells are well suited for this task. The investigation revealed that this fully passive network tends to feedback loops that negatively influence the functionality. Conditions for correct sorting were evaluated and it was found that when cells are preset to high resistive states, the negative loops are avoided. Furthermore, transmission gates were introduced in between the stages of the sorting network which stabilize the system. The transmission gates are essential for the sorting network but lead to some circuit overhead. Low SET voltages are recommendable. Since the upper limit for the potential difference between the inputs scales linearly with the reset voltage, the RESET voltages of the cell should not be too small, though.

5 Résumé

In this work the potential of resistive switches for brain inspired and fuzzy logic applications has been highlighted.

5.1 Summary

Minimum and maximum gates for fuzzy logic operations were introduced and demonstrated. The basic concept was proven experimentally on functional micro sized resistively switching devices. Compared to conventional CMOS approaches, the presented ReRAM approach offers smaller unit array and superior scaling properties whereas the power consumption can be kept low or even be reduced.

Whereas ReRAM devices enable energy-efficient operations in principle [71], the question of whether the energy-efficiency of ReRAM-type MIN/MAX gate-based circuits is better than comparable CMOS circuit cannot be answered without knowing the area of application (e.g. sorting or audio signal processing) and actual circuit implementation. Certainly, further improvements of ReRAM cell performance in terms of reliability, cycle-to-cycle variance and endurance are required to enable ReRAM based memory and logic applications.

It was disclosed that the presented MIN and MAX logic gates offer limited performance in terms of interconnectivity. Since there is no active signal restoration within the passive memristive device, additional amplification circuits between the stages is needed. One option is an add-on of analog buffer stages in the circuitry which recover the redirected signal. The high input and low output impedance of those stages would also have a positive effect on the system. However, it was also shown that this system can work adequately enough in small circuit blocks without buffers, for example as an implementation of area and energy efficient memristive sorting networks.

AgI-type ECM devices were taken as the core element for sorting network simulations. They turned out to be suitable for this task when input voltages are not too large. Their benefit is that input voltage differences can be rather small. Voltages in the range of some millivolts to volts are acceptable. However, every sorting network requires an individual adaption of the chosen material system, with respect of SET and RESET voltages, maximum input voltages, R_{off}/R_{on} ratio and switching times.

Based on the complementary assembly of two RRAM cells in combination with a non-destructive readout mechanism an associative memory was introduced and explained. It was demonstrated that ACNs are capable to detect Hamming distances between search and stored patterns. This feature goes beyond the typical properties of conventional content addressable memories. It does not require CMOS integration and can be integrated in fully passive circuitry. In addition to that, the information stored is non-volatile and thus does not require continuous energy consumption.

Feasibility of reprogrammable associative capacitive networks was proven by simulative and physical experiments. Valid tests were performed in the nanosecond regime on suitable micro-structures. By accurate dynamical simulations ECM-type ACN devices were studied for the technologically relevant feature size of 40 nm. By considering realistic parasitic capacitances and line resistances, it can be shown that the energy consumption of a 32 bit ACN array is less than 0.01 fJ/bit/search. In addition, simulations on a more abstract device model could show that integrations of 41×40 large arrays lead to meaningful results in terms of similarity detection.

Brain-inspired computation was implemented by the idea of ACNs and reveals a lot of new possibilities on the upcoming path towards future neuromorphic computers. The architecture of ACNs can be used as a memory-intensive platform for in memory computation approaches.

5.2   Outlook

Neuromorphic computation is considered as one of the most important future topics in nanoelectronics. The successful demonstration of a brain-inspired associative capacitive network capable of similarity detection can be an important key-enabler for upcoming tasks like pattern recognition. This task is not only limited to the herein presented content addressable memory application but opens doors for future in memory computation tasks. As postulated in [29], the bottle neck between memory and computation unit can be overcome by the ability to perform computer operations directly within the memory. ACNs can be an important piece to achieve this demand.

With the ongoing technological progress in the field of sensorics and big data processing, studies of fuzzy logic will certainly regain importance. Direct operations on non-digital information will become more and more relevant. That is also endorsed by the multi-state property of resistively switching devices.

List of Figures

List of Tables

Bibliography

[1]   ITRS, "The International Technology Roadmap for Semiconductors 2.0," *ITRS - 2015 Edition*, 2015.

[2]   G. E. Moore, "Cramming more components onto integrated circuits," *Electronics*, vol. 38, pp. 114-117, 1965.

[3]   N. Z. Haron and S. Hamdioui, "Why is CMOS scaling coming to an END?," *2008 3rd International Design and Test Workshop*, vol., pp. 98-103, 2008.

[4]   IRDS, "The International Roadmap for Devices and Systems," *IRDS - 2017 Edition*, 2017.

[5]   R. Waser, R. Dittmann, G. Staikov and K. Szot, "Redox-Based Resistive Switching Memories - Nanoionic Mechanisms, Prospects, and Challenges," *Adv. Mater.*, vol. 21, pp. 2632-2663, 2009.

[6]   D. B. Strukov, G. S. Snider, D. R. Stewart and R. S. Williams, "The missing memristor found," *Nature*, vol. 453, pp. 80-83, 2008.

[7]   L.O. Chua and S.M. Kang, "Memristive devices and systems," *Proc. IEEE*, vol. 64, pp. 209-223, 1976.

[8]   L.O. Chua, "Resistance switching memories are memristors," *Appl. Phys. A-Mater. Sci. Process.*, vol. 102, pp. 765-783, 2011.

[9]   E. Linn, R. Rosezin, S. Tappertzhofen, U. Böttger and R. Waser, "Beyond von Neumann-logic operations in passive crossbar arrays alongside memory operations," *Nanotechnology*, vol. 23, pp. 305205, 2012.

[10]   M. Di Ventra and Y. V. Pershin, "The parallel approach," *Nat. Phys.*, vol. 9, pp. 200-202, 2013.

[11]   T. Ohno, T. Hasegawa, T. Tsuruoka, K. Terabe, J. K. Gimzewski and M. Aono, "Short-term plasticity and long-term potentiation mimicked in single inorganic synapses," *Nat. Mater.*, vol. 10, pp. 591-595, 2011.

[12]   T. Shibata, "Computing based on the physics of nano devices-A beyond-CMOS approach to human-like intelligent systems," *Solid-State Electron.*, vol. 53, pp. 1227-1241, 2009.

[13]   T. Shibata, R. Zhang, S. P. Levitan, D. E. Nikonov and G. I. Bourianoff, "CMOS supporting circuitries for nano-oscillator-based associative memories," *2012 13th*

International Workshop On Cellular Nanoscale Networks and Their Applications (cnna 2012), vol., pp. 1-5, 2012.

[14] M. Verleysen, D. Martin and P. Jespers, "A capacitive neural network for associative memory," *Proceedings of the 10th Symposium on Information Theory in the Benelux*, vol., pp. 73-79, 1989.

[15] K. Pagiamtzis and A. Sheikholeslami, "Content-addressable memory (CAM) circuits and architectures: A tutorial and survey," *IEEE J. Solid-State Circuits*, vol. 41, pp. 712-727, 2006.

[16] A. Goel and P. Gupta, "Small Subset Queries and Bloom Filters Using Ternary Associative Memories, with Applications," *Sigmetrics 2010: Proceedings of the 2010 Acm Sigmetrics International Conference On Measurement and Modeling of Computer Systems*, vol. 38, pp. 143-154, 2010.

[17] H. Noda, K. Inoue, M. Kuroiwa, F. Igaue, K. Yamamoto, H. Mattausch, T. Koide, A. Amo, A. Hachisuka, S. Soeda, I. Hayashi, F. Morishita, K. Dosaka, K. Arimoto, K. Fujishimia, K. Anami and T. Yoshihara, "A cost-efficient high-performance dynamic TCAM with pipelined hierarchical searching and shift redundancy architecture," *IEEE J. Solid-State Circuits*, vol. 40, pp. 245-253, 2005.

[18] T. Hanyu, N. Kanagawa and M. Kameyama, "Non-volatile one-transistor-cell multiple-valued CAM with a digit-parallel-access scheme and its applications," *Computers & Electrical Engineering*, vol. 23, pp. 407-414, 1997.

[19] W. Xu, T. Zhang and Y. Chen, "Design of Spin-Torque Transfer Magnetoresistive RAM and CAM/TCAM with High Sensing and Search Speed," *IEEE Transactions on Very Large Scale Integration (VLSI) Systems*, vol. 18, pp. 66-74, 2010.

[20] S. Matsunaga, K. Hiyama, A. Matsumoto, S. Ikeda, H. Hasegawa, K. Miura, J. Hayakawa, T. Endoh, H. Ohno and T. Hanyu, "Standby-Power-Free Compact Ternary Content-Addressable Memory Cell Chip Using Magnetic Tunnel Junction Devices," *Appl. Phys. Express*, vol. 2, pp. 23004/1-3, 2009.

[21] S. Matsunaga, T. Hanyu, H. Kimura, T. Nakamura and H. Takasu, "Implementation of a standby-power-free CAM based on complementary ferroelectric-capacitor logic," *2007 Asia and South Pacific Design Automation Conference*, pp. 116-117, 2007.

[22] L. Zheng, S. Shin and S. M. S. Kang, "Memristor-based ternary content addressable memory (mTCAM) for data-intensive computing," *Semicond. Sci. Tech.*, vol. 29, pp. 104010, 2014.

[23] S.-J. Lee, K.-S. Oh, Y.-G. Ahn, K. Cho and K. Eshraghian, "Complementary Resistive Switch (CRS) Based Smart Sensor Search Engine," *2013 IEEE Eighth International Conference on Intelligent Sensors, Sensor Networks and Information Processing*, pp. 485-490, 2013.

[24] L.-Y. Huang, M.-F. Chang, C.-H. Chuang, C.-C. Kuo, C.-F. Chen, G.-H. Yang, H.-J. Tsai, T.-F. Chen, S.-S. Sheu, K.-L. Su, F. T. Chen, T.-K. Ku, M.-J. Tsai and M.-J. Kao, "ReRAM-based 4T2R nonvolatile TCAM with 7x NVM-stress reduction, and 4x improvement in speed-wordlength-capacity for normally-off instant-on filter-based search engines used in big-data processing," *2014 Symposium On Vlsi Circuits Digest of Technical Papers*, pp. 122-123, 2014.

[25] F. Alibart, T. Sherwood and D. Strukov, "Hybrid CMOS/nanodevice circuits for high throughput pattern matching applications," *Proceedings of the 2011 NASA/ESA Conference On Adaptive Hardware and Systems (AHS)* pp. 279-286, 2011.

[26] K. Eshraghian, K. R. Cho, O. Kavehei, S. K. Kang, D. Abbott and S. M. S. Kang, "Memristor MOS Content Addressable Memory (MCAM): Hybrid Architecture for Future High Performance Search Engines," *IEEE Transactions on Very Large Scale Integration (VLSI) Systems*, vol. 19, pp. 1407-1417, 2011.

[27] E. Linn, R. Rosezin, C. Kügeler and R. Waser, "Complementary Resistive Switches for Passive Nanocrossbar Memories," *Nat. Mater.*, vol. 9, pp. 403-406, 2010.

[28] S. H. Jo, T. Chang, I. Ebong, B. B. Bhadviya, P. Mazumder and W. Lu, "Nanoscale Memristor Device as Synapse in Neuromorphic Systems," *Nano Lett.*, vol. 10, pp. 1297-1301, 2010.

[29] S. Hamdioui, L. Xie, H. A. Du Nguyen, M. Taouil, K. Bertels, H. Corporaal, H. Jiao, F. Catthoor, D. Wouters, E. Linn and J. van Lunteren, "Memristor based computation-in-memory architecture for data-intensive applications," *2015 Design, Automation Test in Europe Conference Exhibition (DATE)*, pp. 1718-1725, 2015.

[30] John von Neumann, *Theory of Self-Reproducing Automata* University of Illinois Press, 1966.

[31] R. Waser, "Memory Devices and Storage Systems - Introduction to Part V," in *Nanoelectronics and Information Technology*, 3rd ed., Wiley-VCH, pp. 603-620, 2012.

[32] G. Binasch, P. Grünberg, F. Saurenbach and W. Zinn, "Enhanced magnetoresistance in layered magnetic structures with antiferromagnetic interlayer exchange," *Phys. Rev. B: Condens. Matter*, vol. 39, pp. 4828-4830, 1989.

[33] M. N. Baibich, J. M. Broto, A. Fert, F. N. Van Dau, F. Petroff, P. Etienne, G. Creuzet, A. Friederich and J. Chazelas, "Giant Magnetoresistance of (001)Fe/(001)Cr Magnetic Superlattices," *Phys. Rev. Lett.*, vol. 61, pp. 2472-2475, 1988.

[34] F. Masuoka, M. Momodomi, Y. Iwata and R. Shirota, "New ultra high density EPROM and flash EEPROM with NAND structure cell," *1987 International Electron Devices Meeting*, pp. 552-555, 1987.

[35] H. Akinaga, "Recent advances and future prospects in functional-oxide nanoelectronics: The emerging materials and novel functionalities that are accelerating semiconductor device research and development," *Jpn. J. Appl. Phys.*, vol. 52, pp. 100001, 2013.

[36] D. Ielmini and R. Waser, *Resistive Switching - From Fundamentals of Nanoionic Redox Processes to Memristive Device Applications* Wiley-VCH, 2016.

[37] D. Ielmini, R. Bruchhaus and R. Waser, "Thermochemical resistive switching: materials, mechanisms, and scaling projections," *Phase Transit.*, vol. 84, pp. 570-602, 2011.

[38] I. Valov, "Redox-Based Resistive Switching Memories (ReRAMs): Electrochemical Systems at the Atomic Scale," *ChemElectroChem*, vol. 1, pp. 26-36, 2014.

[39] I. Valov, R. Waser, J. R. Jameson and M. N. Kozicki, "Electrochemical metallization memories-fundamentals, applications, prospects," *Nanotechnology*, vol. 22, pp. 254003/1-22, 2011.

[40] I. Valov, I. Sapezanskaia, A. Nayak, T. Tsuruoka, T. Bredow, T. Hasegawa, G. Staikov, M. Aono and R. Waser, "Atomically controlled electrochemical nucleation at superionic solid electrolyte surfaces," *Nat. Mater.*, vol. 11, pp. 530-535, 2012.

[41] I. Valov and G. Staikov, "Nucleation and growth phenomena in nanosized electrochemical systems for resistive switching memories," *J. Solid State Electrochem.*, vol. 17, pp. 365-371, 2013.

[42] I. Valov and M. N. Kozicki, "Cation-based resistance change memory," *J. Phys. D Appl. Phys.*, vol. 46, pp. 074005, 2013.

[43] R. Waser and M. Aono, "Nanoionics-based resistive switching memories," *Nat. Mater.*, vol. 6, pp. 833-840, 2007.

[44] J. J. Yang, D. B. Strukov and D. R. Stewart, "Memristive Devices for Computing," *Nat. Nanotechnol.*, vol. 8, pp. 13-24, 2013.

[45] F. Pan, S. Gao, C. Chen, C. Song and F. Zeng, "Recent progress in resistive random access memories: Materials, switching mechanisms, and performance," *Mater. Sci. Eng. R-Rep.*, vol. 83, pp. 1-59, 2014.

[46] M. N. Kozicki, M. Park and M. Mitkova, "Nanoscale memory elements based on solid-state electrolytes," *IEEE Trans. Nanotechnol.*, vol. 4, pp. 331-338, 2005.

[47] L. Yang, C. Kuegeler, K. Szot, A. Ruediger and R. Waser, "The influence of copper top electrodes on the resistive switching effect in TiO_2 thin films studied by conductive atomic force microscopy," *Appl. Phys. Lett.*, vol. 95, pp. 013109, 2009.

[48] C. Schindler, S. C. P. Thermadam, R. Waser and M. N. Kozicki, "Bipolar and unipolar resistive switching in Cu-doped SiO_2," *IEEE Trans. Electron Devices*, vol. 54, pp. 2762-2768, 2007.

[49] S. Tappertzhofen, M. Hempel, I. Valov and R. Waser, "Proton Mobility in SiO_2 Thin Films and Impact of Hydrogen and Humidity on the Resistive Switching Effect," *Mater. Res. Soc. Symp. Proc.*, vol. 1330, 2011.

[50] C. Schindler, "Resistive switching in electrochemical metallization memory cells," Ph.D. dissertation, RWTH Aachen University, 2009.

[51] K. Szot, M. Rogala, W. Speier, Z. Klusek, A. Besmehn and R. Waser, "TiO_2 - a prototypical memristive material," *Nanotechnology*, vol. 22, pp. 254001/1-21, 2011.

[52] D. S. Jeong, H. Schroeder and R. Waser, "Coexistence of bipolar and unipolar resistive switching behaviors in a $Pt/TiO_2/Pt$ stack," *Electrochem. Solid State Lett.*, vol. 10, pp. G51-G53, 2007.

[53] J. J. Yang, M. D. Pickett, X. Li, D. A. A. Ohlberg, D. R. Stewart and R. S. Williams, "Memristive switching mechanism for metal/oxide/metal nanodevices," *Nat. Nanotechnol.*, vol. 3, pp. 429-433, 2008.

[54] J. Park, K. P. Biju, S. Jung, W. Lee, J. Lee, S. Kim, S. Park, J. Shin and H. Hwang, "Multibit Operation of TiOx-Based ReRAM by Schottky Barrier Height Engineering," *IEEE Electron Device Lett.*, vol. 32, pp. 476-478, 2011.

[55] L. Zhang, Y. Y. Hsu, F. T. Chen, H. Y. Lee, Y. S. Chen, W. S. Chen, P. Y. Gu, W. H. Liu, S. M. Wang, C. H. Tsai, R. Huang and M. J. Tsai, "Experimental investigation of the reliability issue of RRAM based on high resistance state conduction," *Nanotechnology*, vol. 22, pp. 254016, 2011.

[56] P. Chen, Y. Chen, H. Lee, T. Wu, K. Tsai, P. Gu, W. Chen, C. Tsai, F. Chen and M. Tsai, "Impacts of device architecture and low current operation on resistive

switching of HfOx nanoscale devices," *Microelectronic Engineering*, vol. 105, pp. 40-45, 2013.

[57] Z. Wei, Y. Kanzawa, K. Arita, Y. Katoh, K. Kawai, S. Muraoka, S. Mitani, S. Fujii, K. Katayama, M. Iijima, T. Mikawa, T. Ninomiya, R. Miyanaga, Y. Kawashima, K. Tsuji, A. Himeno, T. Okada, R. Azuma, K. Shimakawa, H. Sugaya, T. Takagi, R. Yasuhara, H. Horiba, H. Kumigashira and M. Oshima, "Highly Reliable TaOx ReRAM and Direct Evidence of Redox Reaction Mechanism," *2008 IEEE International Electron Devices Meeting*, pp. 1-4, 2008.

[58] M.-J. Lee, C. B. Lee, D. Lee, S. R. Lee, M. Chang, J. H. Hur, Y.-B. Kim, C. -J. Kim, D. H. Seo, S. Seo, U.-I. Chung, I.-K. Yoo and K. Kim, "A fast, high-endurance and scalable non-volatile memory device made from asymmetric Ta_2O_{5-x}/TaO_{2-x} bilayer structures," *Nat. Mater.*, vol. 10, pp. 625-630, 2011.

[59] K. Szot, G. Bihlmayer and W. Speier, "Nature of the Resistive Switching Phenomena in TiO_2 and $SrTiO_3$: Origin of the Reversible Insulator–Metal Transition," *Solid State Physics*, vol. 65, pp. 353-559, 2014.

[60] C. Rodenbücher, "Resistive switching phenomena of extended defects in Nb-doped SrTiO3 under influence of external gradients," Ph.D. dissertation, RWTH Aachen University, 2014.

[61] A. Beck, J. G. Bednorz, C. Gerber, C. Rossel and D. Widmer, "Reproducible switching effect in thin oxide films for memory applications," *Appl. Phys. Lett.*, vol. 77, pp. 139-41, 2000.

[62] S. Q. Liu, N. J. Wu and A. Ignatiev, "Electric-pulse-induced reversible resistance change effect in magnetoresistive films," *Appl. Phys. Lett.*, vol. 76, pp. 2749-51, 2000.

[63] R. Waser, R. Bruchhaus and S. Menzel, "Redox-Based Resistive Random Access Memories", in *Nanoelectronics and Information Technology,* 3rd ed., Wiley-VCH, 2012.

[64] S. Menzel, M. Waters, A. Marchewka, U. Böttger, R. Dittmann and R. Waser, "Origin of the Ultra-nonlinear Switching Kinetics in Oxide-Based Resistive Switches," *Adv. Funct. Mater.*, vol. 21, pp. 4487-4492, 2011.

[65] K. Fleck, U. Böttger, R. Waser and S. Menzel, "Interrelation of Sweep and Pulse Analysis of the SET Process in SrTiO3 Resistive Switching Memories," *IEEE Electron Device Lett.*, vol. 35, pp. 924-926, 2014.

[66] R. Meyer, L. Schloss, J. Brewer, R. Lambertson, W. Kinney, J. Sanchez and D. Rinerson, "Oxide Dual-Layer Memory Element for Scalable Non-Volatile Cross-Point Memory Technology," *Proc. NVMTS*, pp. 54-58, 2008.

[67] D. B. Strukov and R. S. Williams, "Exponential ionic drift: fast switching and low volatility of thin-film memristors," *Appl. Phys. A-Mater. Sci. Process.*, vol. 94, pp. 515-519, 2009.

[68] B. Gao, S. Yu, N. Xu, L. F. Liu, B. Sun, X. Y. Liu, R. Q. Han, J. F. Kang, B. Yu and Y. Y. Wang, "Oxide-Based RRAM Switching Mechanism: A New Ion-Transport-Recombination Model," *IEEE International Electron Devices Meeting 2008*, pp. 563-566, 2008.

[69] S. Li, H. Z. Zeng, S. Y. Zhang and X. H. Wei, "Bipolar resistive switching behavior with high ON/OFF ratio of Co:BaTiO3 films by acceptor doping," *Appl. Phys. Lett.*, vol. 102, pp. 153506, 2013.

[70] A. Bricalli, E. Ambrosi, M. Laudato, M. Maestro, R. Rodriguez and D. Ielmini, "SiOx-based resistive switching memory (RRAM) for crossbar storage/select elements with high on/off ratio," *2016 IEEE International Electron Devices Meeting (IEDM)*, pp. 4.3.1-4.3.4, 2016.

[71] R. Waser, V. Rana, S. Menzel and E. Linn, "Energy-efficient Redox-based Non-Volatile Memory Devices and Logic Circuits," *3rd Berkeley Symposium on Energy Efficient Electronic Systems*, pp. 1-2, 2013.

[72] J. van den Hurk, V. Havel, E. Linn, R. Waser and I. Valov, "Ag/GeS$_x$/Pt-based complementary resistive switches for hybrid CMOS/Nanoelectronic logic and memory architectures," *Scientific Reports*, vol. 3, pp. 2856, 2013.

[73] E. Linn, "Complementary Resistive Switches," Ph.D. dissertation, RWTH Aachen University, 2012.

[74] T. Breuer, A. Siemon, E. Linn, S. Menzel, R. Waser and V. Rana, "Low-current operations in $4F^2$-compatible Ta2O5 -based complementary resistive switches," *Nanotechnology*, vol. 26, pp. 415202, 2015.

[75] L. Nielen, "Fabrication, Electrical Characterization and Macro-Modeling of Complementary Resistive Switches (CRS)," Diploma thesis, RWTH Aachen University, 2011.

[76] R. Rosezin, E. Linn, L. Nielen, C. Kügeler, R. Bruchhaus and R. Waser, "Integrated Complementary Resistive Switches for Passive High-Density Nanocrossbar Arrays," *IEEE Electron Device Lett.*, vol. 32, pp. 191-193, 2011.

[77] T. Breuer, "Development of ReRAM-based Devices for Logic- and Computation-in-Memory Applications," Ph.D. dissertation, RWTH Aachen University, 2017.

[78] J. Mustafa, "Design and Analysis of Future Memories Based on Switchable Resistive Elements," Ph.D. dissertation, RWTH Aachen University, 2006.

[79] S. Tappertzhofen, E. Linn, L. Nielen, R. Rosezin, F. Lentz, R. Bruchhaus, I. Valov, U. Böttger and R. Waser, "Capacity based Nondestructive Readout for Complementary Resistive Switches," *Nanotechnology*, vol. 22, pp. 395203/1-7, 2011.

[80] X. T. Zhang, Q. X. Yu, Y. P. Yao and X. G. Li, "Ultrafast resistive switching in SrTiO$_3$:Nb single crystal," *Appl. Phys. Lett.*, vol. 97, pp. 222117/1-3, 2010.

[81] C. Yoshida, K. Tsunoda, H. Noshiro and Y. Sugiyama, "High speed resistive switching in Pt/TiO$_2$/TiN film for nonvolatile memory application," *Appl. Phys. Lett.*, vol. 91, pp. 223510/1-3, 2007.

[82] C. Kuegeler, C. Nauenheim, M. Meier, R. Ruediger and R. Waser, "Fast resistance switching of TiO$_2$ and MSQ thin films for non-volatile memory applications (RRAM)," *2008 9th Annual Non-Volatile Memory Technology Symposium (NVMTS)*, pp. 59-63, 2008.

[83] Y. Yang, P. Gao, S. Gaba, T. Chang, X. Pan and W. Lu, "Observation of conducting filament growth in nanoscale resistive memories," *Nature Communications*, vol. 3, pp. 732, 2012.

[84] Y. Yang, P. Gao, L. Li, X. Pan, S. Tappertzhofen, S. Choi, R. Waser, I. Valov and W. D. Lu, "Electrochemical dynamics of nanoscale metallic inclusions in dielectrics," *Nat. Commun.*, vol. 5, pp. 4232/1-9, 2014.

[85] L. Yang, C. Kügeler, K. Szot, A. Rüdiger and R. Waser, "The influence of copper top electrodes on the resistive switching effect in TiO$_2$ thin films studied by conductive atomic force microscopy," *Appl. Phys. Lett.*, vol. 95, pp. 13109, 2009.

[86] P. Alexandrov, J. Koprinarova and D. Todorov, "Dielectric properties of TiO$_2$-films reactively sputtered from Ti in an RF magnetron," *Vacuum*, pp. 1333-1336, 1996.

[87] H. R. Philipp, "Optical properties of non-crystalline Si, SiO, SiO$_x$ and SiO$_2$," *J. Phys. Chem. Sol.*, vol. 32, pp. 1935-1945, 1971.

[88] K. Terabe, T. Hasegawa, T. Nakayama and M. Aono, "Quantized conductance atomic switch," *Nature*, vol. 433, pp. 47-50, 2005.

[89] C. Schindler, G. Staikov and R. Waser, "Electrode kinetics of Cu-SiO$_2$-based resistive switching cells: Overcoming the voltage-time dilemma of electrochemical metallization memories," *Appl. Phys. Lett.*, vol. 94, pp. 072109/1-3, 2009.

[90] R. E. Howard, D. B. Schwartz, J. S. Denker, R. W. Epworth, H. P. Graf, W. E. Hubbard, L. D. Jackel, B. L. Straughn and D. M. Tennant, "An associative memory based on an electronic neural network architecture," *IEEE Trans. Electron Devices*, vol. 34, pp. 1553-1556, 1987.

[91] O. Kavehei, E. Linn, L. Nielen, S. Tappertzhofen, S. Skafidas, I. Valov and R. Waser, "Associative Capacitive Network based on Nanoscale Complementary Resistive Switches for Memory-Intensive Computing," *Nanoscale*, vol. 5, pp. 5119-5128, 2013.

[92] H. J. Mattausch, W. Imafuku, A. Kawabata, T. Ansari, M. Yasuda and T. Koide, "Associative Memory for Nearest-Hamming-Distance Search Based on Frequency Mapping," *IEEE J. Solid-State Circuits*, vol. 47, pp. 1448-1459, 2012.

[93] M. Ikeda and K. Asada, "Time-domain minimum-distance detector and its application to low power coding scheme on chip interface," *Proceedings of the 24th European Solid-State Circuits Conference*, pp. 464-467, 1998.

[94] H. J. Mattausch, T. Gyohten, Y. Soda and T. Koide, "Compact associative-memory architecture with fully parallel search capability for the minimum Hamming distance," *IEEE J. Solid-State Circuits*, vol. 37, pp. 218-227, 2002.

[95] Y. Yano, T. Koide and H. J. Mattausch, "Associative memory with fully parallel nearest-manhattan-distance search for low-power real-time single-chip applications,", 2004, pp. 543-544.

[96] Y. Oike, M. Ikeda and K. Asada, "A high-speed and low-voltage associative co-processor with exact Hamming/Manhattan-distance estimation using word-parallel and hierarchical search architecture," *IEEE J. Solid-State Circuits*, vol. 39, pp. 1383-1387, 2004.

[97] S. Nakahara and T. Kawata, "A design for a minimum Hamming-distance search using asynchronous digital techniques," *IEEE J. Solid-State Circuits*, vol. 40, pp. 276-285, 2005.

[98] W. McCulloch and W. Pitts, "A logical calculus of the ideas immanent in nervous activity," *Bull. Math. Biophys.*, vol. 5, pp. 115-133, 1943.

[99] J. G. Brennan, "A Handbook of Logic," *Philosophical Quarterly*, vol. 12, pp. 280-281, 1962.

[100] Victor P. Nelson, H. Troy Nagle, J. David Irwin and Bill D. Carroll, *Digital Logic Circuit Analysis and Design* Pearson, 1995.

[101] T. Miwa, H. Yamada, Y. Hirota, T. Satoh and H. Hara, "A 1-Mb 2-Tr/b nonvolatile CAM based on flash memory technologies," *IEEE J. Solid-State Circuits*, vol. 31, pp. 1601-1609, 1996.

[102] S. Menzel, U. Böttger and R. Waser, "Simulation of multilevel switching in electrochemical metallization memory cells," *J. Appl. Phys.*, vol. 111, pp. 014501/1-5, 2012.

[103] E. Linn, S. Menzel, S. Ferch and R. Waser, "Compact modeling of CRS devices based on ECM cells for memory, logic and neuromorphic applications," *Nanotechnology*, vol. 24, pp. 384008, 2013.

[104] E. Linn, S. Menzel, R. Rosezin, U. Böttger, R. Bruchhaus and R. Waser, "Modeling Complementary Resistive Switches by Nonlinear Memristive Systems," *Proceedings of the 11th IEEE Conference on Nanotechnology*, pp. 1474-1478, 2011.

[105] O. Kavehei, S. Al-Sarawi, K. R. Cho, K. Eshraghian and D. Abbott, "An Analytical Approach for Memristive Nanoarchitectures," *IEEE Trans. Nanotechnol.*, vol. 11, pp. 374-385, 2012.

[106] *Renesas Electronics: 20Mbit Quad-Search Content-Addressable Memory*, Renesas, 2015.

[107] L. Nielen, S. Tappertzhofen, E. Linn, R. Waser and O. Kavehei, "An Experimental Associative Capacitive Network based on Complementary Resistive Switches for Memory-intensive Computing," *2014 IEEE Silicon Nanoelectronics Workshop*, pp. 1-2, 2014.

[108] L. Nielen, S. Tappertzhofen, E. Linn, O. Kavehei, S. Skafidas, I. Valov and R. Waser, "Live demonstration: An associative capacitive network based on nanoscale complementary resistive switches," *IEEE International Symposium on Circuits and Systems (ISCAS)*, pp. 439, 2014.

[109] J. Choi and B. J. Sheu, "A High-Precision VLSI Winner-Take-All Circuit For Self-Organizin Neural Networks," *IEEE J. Solid-State Circuits*, vol. 28, pp. 576-584, 1993.

[110] F. Yu, T. Lakshman, M. Motoyama and R. Katz, "SSA A Power and Memory Efficient Scheme to Multi-Match Packet Classification," *ACM/IEEE Symposium on Architectures for Networking and Communications Systems*, pp. 105-113, 2005.

[111] K. Lakshminarayanan, A. Rangarajan and S. Venkatachary, "Algorithms for Advanced Packet Classification with Ternary CAMs," *Proceedings of the 2005 conference on Applications, technologies, architectures, and protocols for computer communications*, pp. 193-204, 2005.

[112] J. C. Gallagher, "The Once and Future Analog Alternative: Evolvable Hardware and Analog Computation," *NASA/DoD Conference on Evolvable Hardware*, pp. 43-49, 2003.

[113] F. L. Traversa and M. Di Ventra, "Universal Memcomputing Machines," *IEEE Transactions on Neural Networks and Learning Systems*, vol. 26, pp. 2702-2715, 2015.

[114] M. Prezioso, F. Merrikh-Bayat, B. D. Hoskins, G. C. Adam, K. K. Likharev and D. B. Strukov, "Training and operation of an integrated neuromorphic network based on metal-oxide memristors," *Nature*, vol. 521, pp. 61-64, 2015.

[115] J. Borghetti, G. S. Snider, P. J. Kuekes, J. J. Yang, D. R. Stewart and R. S. Williams, "'Memristive' switches enable 'stateful' logic operations via material implication," *Nature*, vol. 464, pp. 873-876, 2010.

[116] T. Breuer, A. Siemon, E. Linn, S. Menzel, R. Waser and V. Rana, "A HfO2-Based Complementary Switching Crossbar Adder," *Adv. Electron. Mater.*, vol. 1, pp. 1500138, 2015.

[117] P. Kipfer and R. Westermann, *Improved GPU Sorting,* Addison-Wesley, 2005.

[118] R. Marcelino, H. C. Neto and J. M. P. Cardoso, "A Comparison of Three Representative Hardware Sorting Units," *2009 35th Annual Conference of IEEE Industrial Electronics*, pp 2805-2810, 2009.

[119] A. Chattopadhyay, "Ingredients of Adaptability: A Survey of Reconfigurable Processors," *VLSI Design*, vol. 2013, pp. 683615, 2013.

[120] M. Fratrik, M. Klimo, O. Such and O. Skvarek, "Memristive sorting networks," *Phys. Status Solidi C*, vol. 12, pp. 233-237, 2015.

[121] I. Vourkas, D. Stathis and G. Ch. Sirakoulis, "Memristor-based parallel sorting approach using one-dimensional cellular automata," *Electron. Lett.*, vol. 50, pp. 1819-1821, 2014.

[122] M. Klimo and O. Such, "Memristors can implement fuzzy logic," *arXiv:*1110.2074, pp. 1-11, 2011.

[123] O. Šuch, E. Linn, M. Klimo, P. Jancovic, M. Frátrik and K. Fröhlich, "On Passive Permutation Circuits," *IEEE J. Emerging Sel. Top. Circuits Syst.*, vol. 5, pp. 173-182, 2015.

[124] L. A. Zadeh, "Fuzzy Sets," *Information and Control*, vol. 8, pp. 338-353, 1965.

[125] J. M. Mendel, "Fuzzy Logic Systems for Engineering: A Tutorial," *Proc. IEEE*, vol. 83, pp. 345-377, 1995.

[126] R. Waser, "Logic Devices – Introduction to Part III," in *Nanoelectronics and Information Technology,* 3rd ed., Wiley-VCH, pp. 323-340, 2012.

[127] S. Kvatinsky, N. Wald, G. Satat, A. Kolodny, U. Weiser and E. Friedman, "MRL - Memristor Ratioed Logic," *2012 13th International Workshop on Cellular Nanoscale Networks and their Applications (CNNA)*, pp. 1-6, 2012.

[128] L. Nielen, S. Ohm, O. Such, M. Klimo, R. Waser and E. Linn, "Memristive Sorting Networks Enabled by Electrochemical Metallization Cells," *Int. J. Unconv. Comput.*, vol. 12, pp. 303-317, 2016.

[129] M. Tada, T. Sakamoto, M. Miyamura, N. Banno, K. Okamoto, N. Iguchi, T. Nohisa and H. Hada, "Highly reliable, complementary atom switch (CAS) with low programming voltage embedded in Cu BEOL for Nonvolatile Programmable Logic," *2011 International Electron Devices Meeting*, pp. 30.2.1-30.2.4, 2011.

[130] T. Breuer, L. Nielen, B. Roesgen, R. Waser, V. Rana and E. Linn, "Realization of Minimum and Maximum Gate Function in Ta_2O_5-based Memristive Devices," *Sci Rep*, vol. 6, pp. 23967/1-9, 2016.

[131] A. C. Torrezan, J. P. Strachan, G. Medeiros-Ribeiro and R. S. Williams, "Sub-nanosecond switching of a tantalum oxide memristor," *Nanotechnology*, vol. 22, pp. 485203, 2011.

[132] T. Tamura, T. Hasegawa, K. Terabe, T. Nakayama, T. Sakamoto, H. Sunamura, H. Kawaura, S. Hosaka and M. Aono, "Switching property of atomic switch controlled by solid electrochemical reaction," *Jpn. J. Appl. Phys.*, vol. 45, pp. L364-L366, 2006.

[133] B. Govoreanu, G. S. Kar, Y-Y. Chen, V. Paraschiv, S. Kubicek, A. Fantini, I. P. Radu, L. Goux, S. Clima, R. Degraeve, N. Jossart, O. Richard, T. Vandeweyer, K. Seo, P. Hendrickx, G. Pourtois, H. Bender, L. Altimime, D. J. Wouters, J. A. Kittl and M. Jurczak, "10x10 nm2 Hf/HfOx Crossbar Resistive RAM with Excellent Performance, Reliability and Low-Energy Operation," *2011 International Electron Devices Meeting*, pp. 31.6.1-31.6.4, 2011.

[134] S. Menzel, M. Salinga, U. Böttger and M. Wimmer, "Physics of the Switching Kinetics in Resistive Memories," *Adv. Funct. Mater.*, vol. 25, pp. 6306-6325, 2015.

[135] P.-E. Gaillardon, D. Sacchetto, S. Bobba, Y. Leblebici and G. De Micheli, "GMS: Generic Memristive Structure for Non-Volatile FPGAs," *2012 IEEE/IFIP 20th International Conference on VLSI and System-on-Chip (VLSI-SoC)*, pp. 94-98, 2012.

[136] K. E. Batcher, "Sorting networks and their applications," *Proceedings of the Spring Joint Computer Conference (AFIPS '68)*, pp. 307-314, 1968.

[137] D. E. Knuth, *Sorting and Searching,* Addison-Wesley, 1998.

[138] S. Menzel, S. Tappertzhofen, R. Waser and I. Valov, "Switching Kinetics of Electrochemical Metallization Memory Cells," *PCCP*, vol. 15, pp. 6945-6952, 2013.

[139] E. Linn, A. Siemon, R. Waser and S. Menzel, "Applicability of Well-Established Memristive Models for Simulations of Resistive Switching Devices," *IEEE*

Transactions on Circuits and Systems - Part I: Regular Papers (TCAS-I), vol. 61, pp. 2402-2410, 2014.

[140] Víctor M. Jiménez-Fernández, Ana D. Martínez, Joel Ramírez, Jesús S. Orea, Omar Alba, Pedro Julián, Juan A. Rodríguez, Osvaldo Agamennoni and Omar D. Lifschitz, "VLSI Design of Sorting Networks in CMOS Technology," in *VLSI Design*, E. Tlelo-Cuautle, Ed. Online: IntechOpen, pp. 93-110, 2012.

[141] C. J. Kuo and Z. W. Huang, "Modified odd-even merge-sort network for arbitrary number of inputs," *2001 IEEE International Conference on Multimedia and Expo*, pp. 929-932, 2001.

[142] F. Shi, Z. Yan and M. Wagh, "An Enhanced Multiway Sorting Network Based on n-Sorters," *2014 IEEE Global Conference On Signal and Information Processing*, pp. 60-64, 2014.

[143] R. Marcelino, H. Neto and J. M. P. Cardoso, "Sorting units for FPGA-based embedded systems," *Distributed Embedded Systems: Design, Middleware and Resources*, pp. 11-22, 2008.

[144] T. Talaska and R. Dlugosz, "Analog Sorting Circuit for the Application in Self-Organizing Neural Networks Based on Neural Gas Learning Algorithm," *2015 22nd International Conference Mixed Design of Integrated Circuits & Systems*, pp. 282-286, 2015.

[145] Ciaran J. Brennan and Mukesh Kumar, "High voltage maximum voltage selector circuit with no quiescent current," *United States Patent 9306552*, 2016.

[146] Roland Priemer and Thomas S. Dranger, "Analog voltage maximum selection and sorting circuits," *United States Patent 6188251*, 2001.

[147] J. Wang, "Analysis and Design of an Analog Sorting Network," *IEEE Transactions On Neural Networks*, vol. 6, pp. 962-971, 1995.

[148] S. Rovetta and R. Zunino, "Minimal-connectivity circuit for analogue sorting," *IEE Proc.-Circuit Device Syst.*, vol. 146, pp. 108-110, 1999.

Appendix

Appendix 1

Case	PRE	Stored states	Applied pattern	Match line	Time for rising voltage ramp
A	GND	'1' = C_{low} '0' = C_{high}	'1' = V_{DD} '0' = GND	low V = low HD high V = high HD	Match = 1st result
B	GND	'1' = C_{high} '0' = C_{low}	'1' = V_{DD} '0' = GND	low V = high HD high V = low HD	Match = last result
C	GND	'1' = C_{low} '0' = C_{high}	'1' = GND '0' = V_{DD}	low V = high HD high V = low HD	Match = last result
D	GND	'1' = C_{high} '0' = C_{low}	'1' = GND '0' = V_{DD}	low V = low HD high V = high HD	Match = 1st result
E	V_{DD}	'1' = C_{low} '0' = C_{high}	'1' = V_{DD} '0' = GND	low V = high HD high V = low HD	Match = last result
F	V_{DD}	'1' = C_{high} '0' = C_{low}	'1' = V_{DD} '0' = GND	low V = low HD high V = high HD	Match = 1st result
G	V_{DD}	'1' = C_{low} '0' = C_{high}	'1' = GND '0' = V_{DD}	low V = low HD high V = high HD	Match = 1st result
H	V_{DD}	'1' = C_{high} '0' = C_{low}	'1' = GND '0' = V_{DD}	low V = high HD high V = low HD	Match = last result

Table A.1: Complete list of ACN voltage schemes, cf. Section 3.1.3.

Appendix 2

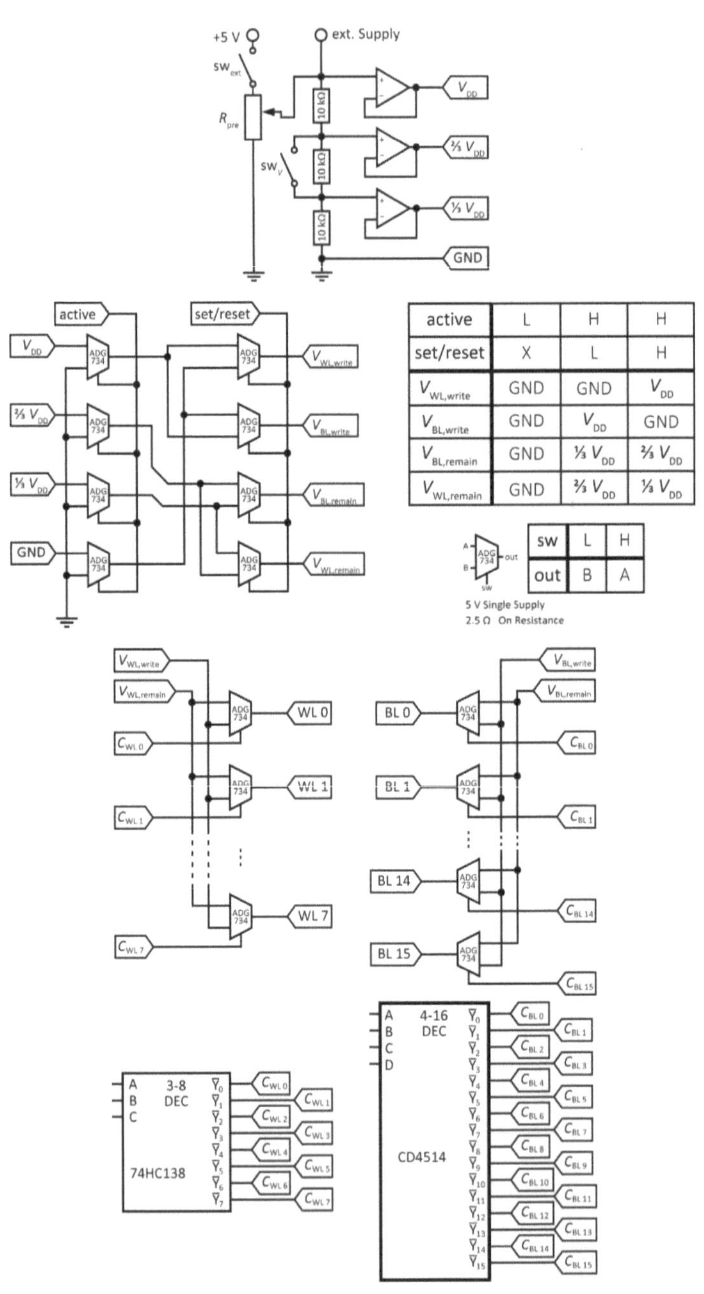

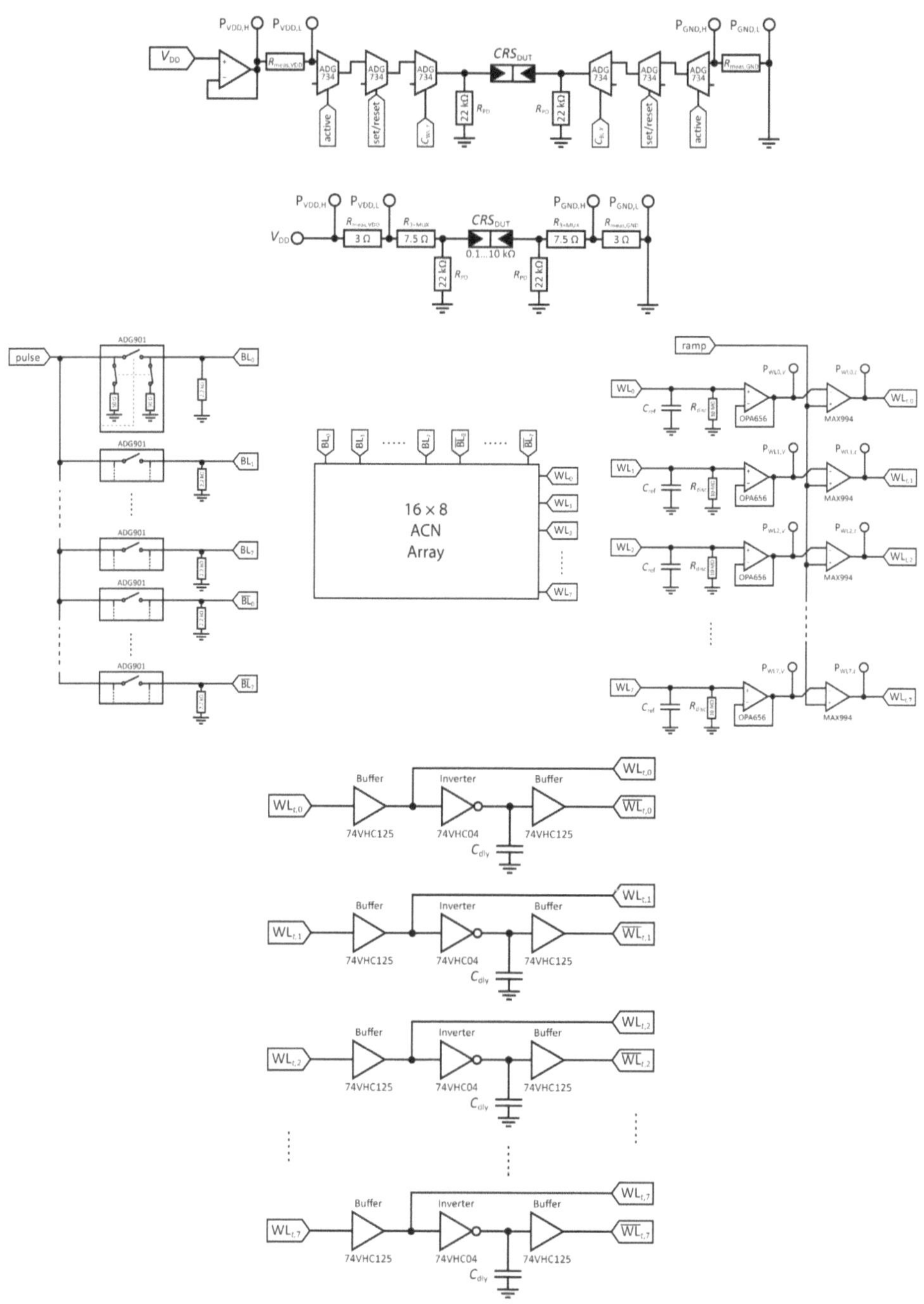

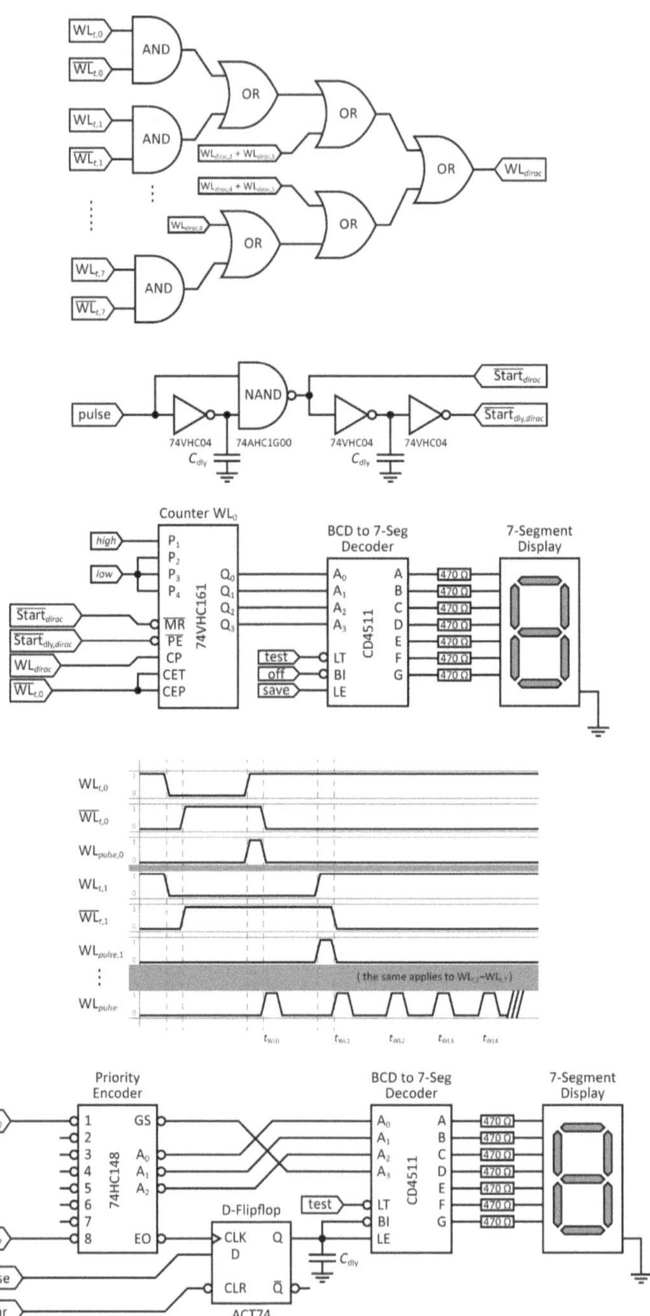

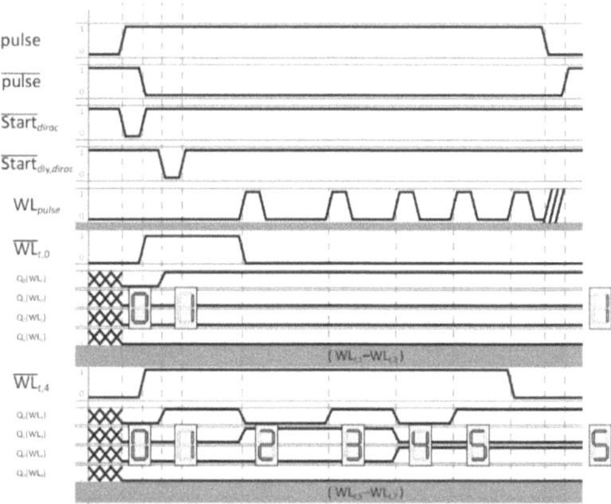

Figure A.1: Schematics and block diagrams for advanced ACN programming and evaluation circuits

Curriculum Vitae

Lutz Nielen

Geburtsdatum: 01.06.1984
Geburtsort: Neuss
Familienstand: verheiratet
Staatsangehörigkeit: deutsch

Bildungsweg

1990 – 1994   Görres-Grundschule, Neuss

1994 – 2003   Alexander-von-Humboldt-Gymnasium, Neuss

2004 – 2010   Studium der Elektrotechnik und Informationstechnik an der RWTH
              Aachen
              Abschluss: Diplom-Ingenieur

Berufliche Tätigkeit

2011   EDS-Design Engineer für Kfz-Bordnetze
              Yazaki Europe Ltd., Köln
              Ford Engineering, Entwicklung

2011-2017   Wissenschaftlicher Angestellter mit Ziel der Promotion
              Institut für Werkstoffe der Elektrotechnik II
              RWTH Aachen

2017-heute   Manager R&D Division
              aixACCT Systems GmbH
              Aachen

List of own publications

Parts of the here presented work have been published in international journals or conferences. The amount of contribution to this work has been discussed and was agreed on with the co-authors.

T. Breuer, L. Nielen, B. Roesgen, R. Waser, V. Rana and E. Linn, "Realization of Minimum and Maximum Gate Function in Ta2O5-based Memristive Devices," Sci Rep, vol. 6, pp. 23967/1-9, 2016.

L. Nielen, A. Siemon, S. Tappertzhofen, R. Waser, S. Menzel and E. Linn, "Complementary resistive switch based neuromorphic associative capacitive network," Nanoelectronic Days, 27-30 April, Juelich, 2015.

L. Nielen, S. Ohm, O. Such, M. Klimo, R. Waser and E. Linn, "Memristive Sorting Networks Enabled by Electrochemical Metallization Cells," Int. J. Unconv. Comput., vol. 12, pp. 303-317, 2016.

L. Nielen, A. Siemon, S. Tappertzhofen, R. Waser, S. Menzel and E. Linn, "Study of Memristive Associative Capacitive Networks for CAM Applications," IEEE J. Emerging Sel. Top. Circuits Syst., vol. 5, pp. 153-161, 2015.

L. Nielen, A. Siemon, S. Tappertzhofen, R. Waser, S. Menzel and E. Linn, "Memristive Associative Capacitive Networks – A Novel Approach for Fully Parallel Pattern Recognition," Workshop on Memristor Technology, Design, Automation and Computing (MemTDAC), 2015.

L. Nielen, S. Tappertzhofen, E. Linn, R. Waser and O. Kavehei, "An Experimental Associative Capacitive Network based on Complementary Resistive Switches for Memory-intensive Computing," 2014 IEEE Silicon Nanoelectronics Workshop, 2014.

L. Nielen, S. Tappertzhofen, E. Linn, O. Kavehei, S. Skafidas, I. Valov and R. Waser, "Live demonstration: An associative capacitive network based on nanoscale complementary resistive switches," IEEE International Symposium on Circuits and Systems (ISCAS), pp. 439, 2014.

O. Kavehei, E. Linn, L. Nielen, S. Tappertzhofen, S. Skafidas, I. Valov and R. Waser, "Associative Capacitive Network based on Nanoscale Complementary Resistive Switches for Memory-Intensive Computing," Nanoscale, vol. 5, pp. 5119-5128, 2013.

R. Rosezin, E. Linn, L. Nielen, C. Kügeler, R. Bruchhaus and R. Waser, "Complementary Resistive Switches (CRS): High speed performance for passive nano-crossbar arrays," Mater. Res. Soc. Symp. Proc., vol. 1337, 2011.

R. Rosezin, E. Linn, <u>L. Nielen</u>, C. Kügeler, R. Bruchhaus and R. Waser, "Inte-grated Complementary Resistive Switches for Passive High-Density Nanocrossbar Arrays," IEEE Electron Device Lett., vol. 32, pp. 191-193, 2011.

S. Tappertzhofen, E. Linn, <u>L. Nielen</u>, R. Rosezin, F. Lentz, R. Bruchhaus, I. Valov, U. Böttger and R. Waser, "Capacity based Nondestructive Readout for Complementary Resistive Switches," Nanotechnology, vol. 22, pp. 395203/1-7, 2011.